T0254686

Multilevel Modeling
Using R
Second Edition

Multilevel Modeling Using R
Second Edition
Second Edition

Second Edition

W. Holmes Finch, Jocelyn E. Bolin,
and Ken Kelley

CRC Press
Taylor & Francis Group
Boca Raton   London   New York

CRC Press is an imprint of the
Taylor & Francis Group, an **informa** business

CRC Press
Taylor & Francis Group
52 Vanderbilt Avenue,
New York, NY 10017

International Standard Book Number-13: 978-1-1384-8071-1 (Hardback)
International Standard Book Number-13: 978-1-1384-8067-4 (Paperback)

Visit the Taylor & Francis Web site at
http://www.taylorandfrancis.com

and the CRC Press Web site at
http://www.crcpress.com

Contents

Authors

W. Holmes Finch is the George and Frances Ball Distinguished Professor of Educational Psychology at Ball State University, where he teaches courses on factor analysis, structural equation modeling, categorical data analysis, regression, multivariate statistics, and measurement to graduate students in psychology and education. Dr. Finch is also an accredited professional statistician (PStat®). He earned a PhD from the University of South Carolina. His research interests include multilevel models, latent variable modeling, methods of prediction and classification, and nonparametric multivariate statistics.

Jocelyn E. Bolin is a professor in the Department of Educational Psychology at Ball State University, where she teaches courses on introductory and intermediate statistics, multiple regression analysis, and multilevel modeling to graduate students in social science disciplines. Dr. Bolin is a member of the American Psychological Association, the American Educational Research Association, and the American Statistical Association, and is an accredited professional statistician (PStat®). She earned a PhD in educational psychology from Indiana University Bloomington. Her research interests include statistical methods for classification and clustering and use of multilevel modeling in the social sciences.

Ken Kelley is the Viola D. Hank associate professor of management in the Mendoza College of Business at the University of Notre Dame. Dr. Kelley is also an accredited professional statistician (PStat®) and associate editor of *Psychological Methods*. His research involves the development, improvement, and evaluation of quantitative methods, especially as they relate to statistical and measurement issues in applied research. He is the developer of the MBESS package for the R statistical language and environment.

1

Linear Models

Statistical models provide powerful tools to researchers in a wide array of disciplines. Such models allow for the examination of relationships among multiple variables, which in turn can lead to a better understanding of the world. For example, sociologists use linear regression to gain insights into how factors such as ethnicity, gender, and level of education are related to an individual's income. Biologists can use the same type of model to understand the interplay between sunlight, rainfall, industrial runoff, and biodiversity in a rain forest. And using linear regression, educational researchers can develop powerful tools for understanding the role that different instructional strategies have on student achievement. In addition to providing a path by which various phenomena can be better understood, statistical models can also be used as predictive tools. For example, econometricians might develop models to predict labor market participation given a set of economic inputs, whereas higher education administrators may use similar types of models to predict grade point average for prospective incoming freshmen in order to identify those who might need academic assistance during their first year of college.

As can be seen from these few examples, statistical modeling is very important across a wide range of fields, providing researchers with tools for both explanation and prediction. Certainly, the most popular of such models over the last 100 years of statistical practice has been the general linear model (GLM). The GLM links a dependent, or outcome variable to one or more independent variables, and can take the form of such popular tools as analysis of variance (ANOVA) and regression. Given its popularity and utility, and the fact that it serves as the foundation for many other models, including the multilevel models featured in this book, we will start with a brief review of the linear model, particularly focusing on regression. This review will include a short technical discussion of linear regression models, followed by a description of how they can be estimated using the R language and environment (R Development Core Team, 2012). The technical aspects of this discussion are purposefully not highly detailed, as we focus on the model from a conceptual perspective. However, sufficient detail is presented so that the reader having only limited familiarity with the linear regression model will be provided with a basis for moving forward to multilevel models, and so that particular features of these more complex models that are shared with the linear model can be explicated. Readers particularly familiar with linear regression and with using R to conduct such analyses may elect to skip this chapter with no loss of understanding in future chapters.

Simple Linear Regression

As noted above, the GLM framework serves as the basis for the multilevel models that we describe in subsequent chapters. Thus, in order to provide the foundation for the rest of the book, we will focus in this chapter on the linear regression model, although its form and function can easily be translated to ANOVA as well. The simple linear regression model in population form is

$$y_i = \beta_0 + \beta_1 x_i + \varepsilon_i, \tag{1.1}$$

where y_i is the dependent variable for individual i in the dataset, and x_i is the independent variable for subject i $(i = 1, \rightleftharpoons, N)$. The terms β_0 and β_1 are the intercept and slope of the model, respectively. In a graphical sense, the intercept is the point where the line in Equation (1.1) crosses the y-axis at $x = 0$. It is also the mean, specifically the conditional mean, of y for individuals with a value of 0 on x, and it is this latter definition that will be most useful in actual practice. The slope, β_1, expresses the relationship between y and x. Positive slope values indicate that larger values of x are associated with correspondingly larger values of y, while negative slopes mean that larger x values are associated with smaller ys. Holding everything else constant, larger values of β_1 (positive or negative) indicate a stronger linear relationship between y and x. Finally, ε_i represents the random error inherent in any statistical model, including regression. It expresses the fact that for any individual i, the model will not generally provide a perfect predicted value of y_i, denoted $\hat{y}_i$ and obtained by applying the regression model as

$$\hat{y}_i = \beta_0 + \beta_1 x_i. \tag{1.2}$$

Conceptually this random error is representative of all factors that might influence the dependent variable other than x.

Estimating Regression Models with Ordinary Least Squares

In virtually all real-world contexts, the population is unavailable to the researcher. Therefore, β_0 and β_1 must be estimated using sample data taken from the population. There exist in the statistical literature several methods for obtaining estimated values of the regression model parameters (b_0 and b_1, respectively) given a set of x and y. By far the most popular and widely used of these methods is ordinary least squares (OLS). The vast majority of other approaches are useful in special cases involving small samples or data that do not conform to the distributional assumptions undergirding OLS. The goal of OLS is to minimize the sum of the squared differences between the observed values of y and the model-predicted values of y, across the sample. This difference, known as the residual, is written as

$$e_i = y_i - \hat{y}_i. \tag{1.3}$$

Therefore, the method of OLS seeks to minimize

$$\sum_{i=1}^{n} e_i^2 = \sum_{i=1}^{n} (y_i - \hat{y}_i)^2. \tag{1.4}$$

The actual mechanism for finding the linear equation that minimizes the sum of squared residuals involves the partial derivatives of the sum of squared function with respect to the model coefficients, β_0 and β_1. We will leave these mathematical details to excellent references, such as Fox (2016). It should be noted that in the context of simple linear regression, the OLS criteria reduce to the following equations, which can be used to obtain b_0 and b_1 as

$$b_1 = r \left(\frac{s_y}{s_x} \right) \tag{1.5}$$

and

$$b_0 = \bar{y} - b_1 \bar{x} \tag{1.6}$$

where

r is the Pearson product moment correlation coefficient between x and y,
s_y is the sample standard deviation of y,
s_x is the sample standard deviation of x,
$\bar{y}$ is the sample mean of y, and
$\bar{x}$ is the sample mean of x.

Distributional Assumptions Underlying Regression

The linear regression model rests upon several assumptions about the distribution of the residuals in the broader population. Although the researcher can typically never collect data from the entire population, it is possible to assess empirically whether these assumptions are likely to hold true based on the sample data. The first assumption that must hold true for linear models to function optimally is that the relationship between y_i and x_i is linear. If the relationship is not linear, then clearly an equation for a line will not provide adequate fit and the model is thus misspecified. A second assumption is that the variance in the residuals is constant regardless of the value of x_i. This assumption is typically referred to as homoscedasticity and is a generalization of the homogeneity of error variance assumption in ANOVA. Homoscedasticity implies that the variance of y_i is constant across values of x_i. The distribution of the dependent variable around the regression line is literally the distribution of the residuals, thus making clear the connection of homoscedasticity of errors with the distribution of y_i around the regression line. The third such assumption is that the residuals are normally distributed in the population. Fourth, it is assumed that the independent variable x is measured without error and that it is unrelated to the model error term, ε. It should be noted that the assumption of x measured without error is not as strenuous as one might first assume. In fact, for most real-world problems, the model will work well even when the independent variable is not error free (Fox, 2016). Fifth and finally, the residuals for any two individuals in the population are assumed to be independent of one another. This independence assumption implies that the unmeasured factors influencing y are not related from one individual to another. It is this assumption that is directly addressed with the use of multilevel models, as we will see in Chapter 2. In many research situations, individuals are sampled in clusters, such that we cannot assume that individuals from the same such cluster will have uncorrelated residuals. For example, if samples are obtained from multiple neighborhoods, individuals within the same neighborhoods may tend to be more like one another than they are like individuals from other neighborhoods. A prototypical example of this is children within schools. Due to a variety of factors, children attending the same school will often have more in common with one another than they do with children from other schools. These "common" things might include neighborhood socioeconomic status, school administration policies, and school learning environment, to name just a few. Ignoring this clustering, or not even realizing it is a problem, can be detrimental to the results of statistical modeling. We explore this issue in great detail later in the book, but for now we simply want to mention that a failure to satisfy the assumption of independent errors is (a) a major problem but (b) often something that can be overcome with the appropriate models, such as multilevel models that explicitly consider the nesting of the data.

Coefficient of Determination

When the linear regression model has been estimated, researchers generally want to measure the relative magnitude of the relationship between the variables. One useful tool for ascertaining the strength of relationship between x and y is the coefficient of determination, which is the squared multiple correlation coefficient, denoted R^2 in the sample. R^2 reflects the proportion of the variation in the dependent variable that is explained by the independent variable. Mathematically, R^2 is calculated as

$$R^2 = \frac{SS_R}{SS_T} = \frac{\sum\limits_{i=1}^{n}\left(\hat{y}_i - \bar{y}\right)^2}{\sum\limits_{i=1}^{n}\left(y_i - \bar{y}\right)^2} = 1 - \frac{\sum\limits_{i=1}^{n}\left(y_i - \hat{y}\right)^2}{\sum\limits_{i=1}^{n}\left(y_i - \bar{y}\right)^2} = 1 - \frac{SS_E}{SS_T}. \tag{1.7}$$

The terms in Equation (1.7) are as defined previously. The value of this statistic always lies between 0 and 1, with larger numbers indicating a stronger linear relationship between x and y, implying that the independent variable is able to account for more variance in the dependent. R^2 is a very commonly used measure of the overall fit of the regression model and, along with the parameter inference discussed below, serves as the primary mechanism by which the relationship between the two variables is quantified.

Inference for Regression Parameters

A second method for understanding the nature of the relationship between x and y involves making inferences about the relationship in the population given the sample regression equation. Because b_0 and b_1 are sample estimates of the population parameters β_0 and β_1, respectively, they are subject to sampling error as is any sample estimate. This means that, although the estimates are unbiased given that the aforementioned assumptions hold, they are not precisely equal to the population parameter values. Furthermore, were we to draw multiple samples from the population and estimate the intercept and slope for each, the values of b_0 and b_1 would differ across samples, even though they would be estimating the same population parameter values for β_0 and β_1. The magnitude of this variation in parameter estimates across samples can be estimated from our single sample using a statistic known as the standard error. The standard error of the slope, denoted as σ_{b_1} in the population, can be thought of as the standard deviation of slope values obtained from all possible samples of size n, taken from the population. Similarly, the standard error of the intercept, σ_{b_0}, is the standard deviation of the intercept values obtained from all such samples. Clearly, it is not possible to obtain census data from a population in an applied research context. Therefore, we will need to estimate the standard errors of both the slope (s_{b_1}) and intercept (s_{b_0}) using data from a single sample, much as we did with b_0 and b_1.

In order to obtain s_{b_1}, we must first calculate the variance of the residuals,

$$s_e^2 = \frac{\sum\limits_{i=1}^{n} e_i^2}{n - p - 1}, \tag{1.8}$$

where

    e_i is the residual value for individual i,

    N is the sample size, and

    p is the number of independent variables (1 in the case of simple regression).

Then

$$s_{b_1} = \frac{1}{\sqrt{1-R^2}} \left[\frac{s_e}{\sqrt{\sum_{i=1}^{n}(x_i - \bar{x})^2}} \right]. \tag{1.9}$$

The standard error of the intercept is calculated as

$$s_{b_0} = s_{b_1} \sqrt{\frac{\sum_{i=1}^{n} x_i^2}{n}}. \tag{1.10}$$

Given that the sample intercept and slope are only estimates of the population parameters, researchers are quite often interested in testing hypotheses to infer whether the data represent a departure from what would be expected in what is commonly referred to as the null case, that the idea of the null value holding true in the population can be rejected. Most frequently (though not always) the inference of interest concerns testing that the population parameter is 0. In particular, a non-0 slope in the population means that x is linearly related to y. Therefore, researchers typically are interested in using the sample to make inferences about whether the population slope is 0 or not. Inferences can also be made regarding the intercept, and again the typical focus is on whether this value is 0 in the population.

    Inferences about regression parameters can be made using confidence intervals and hypothesis tests. Much as with the confidence interval of the mean, the confidence interval of the regression coefficient yields a range of values within which we have some level of confidence (e.g. 95%) that the population parameter value resides. If our particular interest is in whether x is linearly related to y, then we would simply determine whether 0 is in the interval for β_1. If so, then we would not be able to conclude that the population value differs from 0. The absence of a statistically significant result (i.e. an interval not containing 0) does not imply that the null hypothesis is true, but rather it means that there is not sufficient evidence available in the sample data to reject the null. Similarly, we can construct a confidence interval for the intercept, and if 0 is within the interval, we would conclude that the value of y for an individual with $x=0$ could plausibly be, but is not necessarily, 0. The confidence intervals for the slope and intercept take the following forms:

$$b_1 \pm t_{cv} S_{b_1} \tag{1.11}$$

and

$$b_0 \pm t_{cv} S_{b_0} \tag{1.12}$$

Here the parameter estimates and their standard errors are as described previously, while t_{cv} is the critical value of the t distribution for $1-\alpha/2$ (e.g. the .975 quantile if $\alpha=.05$) with $n-p-1$ degrees of freedom. The value of α is equal to 1 minus the desired level of confidence. Thus, for a 95% confidence interval (0.95 level of confidence), α would be 0.05.

In addition to confidence intervals, inference about the regression parameters can also be made using hypothesis tests. In general, the forms of this test for the slope and intercept respectively are

$$t_{b_1} = \frac{b_1 - \beta_1}{S_{b_1}} \tag{1.13}$$

$$t_{b_0} = \frac{b_0 - \beta_0}{S_{b_0}} \tag{1.14}$$

The terms β_1 and β_0 are the parameter values under the null hypothesis. Again, most often the null hypothesis posits that there is no linear relationship between x and y ($\beta_1=0$) and that the value of $y=0$ when $x=0$ ($\beta_0=0$). For simple regression, each of these tests is conducted with $n-2$ degrees of freedom.

Multiple Regression

The linear regression model can very easily be extended to allow for multiple independent variables at once. In the case of two regressors, the model takes the form

$$y_i = \beta_0 + \beta_1 x_{1i} + \beta_2 x_{2i} + \varepsilon_i. \tag{1.15}$$

In many ways, this model is interpreted as is that for simple linear regression. The only major difference between simple and multiple regression interpretation is that each coefficient is interpreted in turn *holding constant* the value of the other regression coefficient. In particular, the parameters are estimated by b_0, b_1, and b_2, and inferences about these parameters are made in the same fashion with regard to both confidence intervals and hypothesis tests. The assumptions underlying this model are also the same as those described for the simple regression model. Despite these similarities, there are three additional topics regarding multiple regression that we need to consider here. These are the inference for the set of model slopes as a whole, an adjusted measure of the coefficient of determination, and the issue of collinearity among the independent variables. Because these issues will be important in the context of multilevel modeling as well, we will address them in detail here.

With respect to model inference, for simple linear regression the most important parameter is generally the slope, so that inference for it will be of primary concern. When there are multiple x variables in the model, the researcher may want to know whether the independent variables taken as a whole are related to y. Therefore, some overall test of model significance is desirable. The null hypothesis for this test is that all of the slopes are equal to 0 in the population; i.e. none of the regressors are linearly related to the dependent variable. The test statistic for this hypothesis is calculated as

$$F = \frac{SS_R / p}{SS_E / (n-p-1)} = \left(\frac{n-p-1}{p}\right)\left(\frac{R^2}{1-R^2}\right). \tag{1.16}$$

Here terms are as defined in Equation (1.7). This test statistic is distributed as an F with p and n-p-1 degrees of freedom. A statistically significant result would indicate that one or more of the regression coefficients are not equal to 0 in the population. Typically, the researcher would then refer to the tests of individual regression parameters, which were described above, in order to identify which were not equal to 0.

A second issue to be considered by researchers in the context of multiple regression is the notion of adjusted R^2. Stated simply, the inclusion of additional independent variables in the regression model will always yield higher values of R^2, even when these variables are not statistically significantly related to the dependent variable. In other words, there is a capitalization on chance that occurs in the calculation of R^2. As a consequence, models including many regressors with negligible relationships with y may produce an R^2 that would suggest the model explains a great deal of variance in y. An option for measuring the variance explained in the dependent variable that accounts for this additional model complexity would be quite helpful to the researcher seeking to understand the true nature of the relationship between the set of independent variables and the dependent. Such a measure exists in the form of the adjusted R^2 value, which is commonly calculated as

$$R_A^2 = 1 - \left(1 - R^2\right)\left(\frac{n-1}{n-p-1}\right). \tag{1.17}$$

R_A^2 only increases with the addition of an x if that x explains more variance than would be expected by chance. R_A^2 will always be less than or equal to the standard R^2. It is generally recommended to use this statistic in practice when models containing many independent variables are used.

A final important issue specific to multiple regression is that of collinearity, which occurs when one independent variable is a linear combination of one or more of the other independent variables. In such a case, regression coefficients and their corresponding standard errors can be quite unstable, resulting in poor inference. It is possible to investigate the presence of collinearity using a statistic known as the variance inflation factor (VIF). In order to obtain the VIF for x_j, we would first regress all of the other independent variables onto x_j and obtain an $R_{x_i}^2$ value. We then calculate

$$\text{VIF} = \frac{1}{1 - R_x^2}. \tag{1.18}$$

The VIF will become large when $R_{x_j}^2$ is near 1, indicating that x_j has very little unique variation when the other independent variables in the model are considered. That is, if the other p-1 regressors can explain a high proportion of x_j, then x_j does not add much to the model, above and beyond the other p-1 regression. Collinearity in turn leads to high sampling variation in b_j, resulting in large standard errors and unstable parameter estimates. Conventional rules of thumb have been proposed for determining when an independent variable is highly collinear with the set of other p-1 regressors. Thus, the researcher might consider collinearity to be a problem if VIF > 5 or 10 (Fox, 2016). The typical response to collinearity is to either remove the offending variable(s) or use an alternative approach to conducting the regression analysis such as ridge regression or regression following a principal components analysis.

Example of Simple Linear Regression by Hand

In order to demonstrate the principles of linear regression discussed above, let us consider a simple scenario in which a researcher has collected data on college grade point average (GPA) and test anxiety using a standard measure where higher scores indicate greater anxiety when taking a test. The sample consisted of 440 college students who were measured on both variables. In this case, the researcher is interested in the extent to which test anxiety is related to college GPA, so that GPA is the dependent and anxiety is the independent variable. The descriptive statistics for each variable, as well as the correlation between the two, appear in Table 1.1.

TABLE 1.1

Descriptive Statistics and Correlation for GPA and Test Anxiety

Variable	Mean	Standard Deviation	Correlation
GPA	3.12	0.51	–0.30
Anxiety	35.14	10.83	

We can use this information to obtain estimates for both the slope and intercept of the regression model using Equations (1.4) and (1.5). First, the slope is calculated as

$$b_1 = -0.30\left(\frac{0.51}{10.83}\right) = -0.014,$$

indicating that individuals with higher test anxiety scores will generally have lower GPAs. Next, we can use this value and information in the table to calculate the intercept estimate: $b_0 = 3.12 - (-0.014)(35.14) = 3.63$.

The resulting estimated regression equation is then $\hat{GPA} = 3.63 - 0.014(\text{Anxiety})$. Thus, this model would predict that for a 1-point increase in the anxiety assessment score, GPA would decrease by –0.014 points.

In order to better understand the strength of the relationship between test anxiety and GPA, we will want to calculate the coefficient of determination. To do this, we need both the SS_R and SS_T, which take the values 10.65 and 115.36, yielding

$$R^2 = \frac{10.65}{115.36} = 0.09.$$

This result suggests that approximately 9% of the variation in GPA is explained by a variation in test anxiety scores. Using this R^2 value and Equation (1.14), we can calculate the F-statistic testing whether any of the model slopes (in this case there is only one) are different from 0 in the population:

$$F = \left(\frac{440-1-1}{1}\right)\left(\frac{0.09}{1-0.09}\right) = 438(0.10) = 43.80.$$

This test has p and n-p-1 degrees of freedom, or 1 and 438 in this situation. The p-value of this test is less than 0.001, leading us to conclude that the slope in the population is indeed significantly different from 0 because the p-value is less than the Type I error rate specified. Thus, test anxiety is linearly related to GPA. The same inference could be conducted using the t-test for the slope. First, we must calculate the standard error of the slope estimate:

$$s_{b_1} = \frac{1}{\sqrt{1-R^2}}\left(\frac{S_E}{\sqrt{\Sigma(x_i - \bar{x})^2}}\right).$$

For these data

$$S_E = \sqrt{\frac{104.71}{440-1-1}} = \sqrt{0.24} = 0.49.$$

In turn, the sum of squared deviations for x (anxiety) was 53743.64, and we previously calculated $R^2 = 0.09$. Thus, the standard error for the slope is

$$s_{b_1} = \frac{1}{\sqrt{1-0.09}} \left(\frac{0.49}{\sqrt{53743.64}} \right) = 1.05(0.002) = 0.002.$$

The test statistic for the null hypothesis that $\beta_1 = 0$ is calculated as

$$t = \frac{b_1 - 0}{s_{b_1}} = \frac{-0.014}{0.002} = -7.00,$$

with n-p-1 or 438 degrees of freedom. The p-value for this test statistic value is less than 0.001 and thus we can probabilistically infer that in the population, the value of the slope is not 0, with the best sample point estimate being −0.014.

Finally, we can also draw inferences about β_1 through a 95% confidence interval, as shown in Equation (1.9). For this calculation, we will need to determine the value of the t distribution with 438 degrees of freedom that corresponds to the 1-0.05/2 or 0.975 point in the distribution. We can do so by using a t table in the back of a textbook, or through standard computer software such as SPSS. In either case, we find that the critical value for this example is 1.97. The confidence interval can then be calculated as

$$(-0.014 - 1.97(0.002), -0.014 + 1.97(0.002))$$
$$(-0.014 - 0.004, -0.104 + 0.004)$$
$$(-0.018, -0.010)$$

The fact that 0 is not in the 95% confidence interval simply supports the conclusion we reached using the p-value as described above. Also, given this interval, we can infer that the actual population slope value lies between −0.018 and −0.010. Thus, anxiety could plausibly have an effect as small as −.01 or as large as −.018.

Regression in R

In R, the function call for fitting linear regression is lm, which is part of the stats library that is loaded by default each time R is started on your computer. The basic form for a linear regression model using lm is:

```
lm(formula, data)
```

Where `formula` defines the linear regression form and `data` indicates the dataset used in the analysis, examples of which appear below. Returning to the previous example, predicting GPA from measures of physical (`BStotal`) and cognitive academic anxiety (`CTA.tot`), the model is defined in R as:

```
Model1.1 <- lm(GPA ~ CTA.tot + BStotal, Cassidy)
```

This line of R code is referred to as a function call, and defines the regression equation. The dependent variable, `GPA`, is followed by the independent variables `CTA.tot` and `BStotal`, separated by ~. The dataset, `Cassidy`, is also given here, after the regression equation has been defined. Finally, the output from this analysis is stored in the object `Model1.1`. In order to view this output, we can type the name of this object in R, and hit return to obtain the following:

```
Call:
lm(formula = GPA ~ CTA.tot + BStotal, data = Cassidy)

Coefficients:
(Intercept)        CTA.tot         BStotal
    3.61892       -0.02007         0.01347
```

The output obtained from the basic function call will return only values for the intercept and slope coefficients, lacking information regarding model fit (e.g. R^2) and significance of model parameters. Further information on our model can be obtained by requesting a summary of the model.

```
summary(Model1.1)
```

Using this call, R will produce the following:

```
Call:
lm(formula = GPA ~ CTA.tot + BStotal, data = Cassidy)

Residuals:
     Min       1Q    Median        3Q       Max
-2.99239 -0.29138   0.01516   0.36849   0.93941

Coefficients:
             Estimate Std. Error t value Pr(>|t|)
(Intercept)  3.618924   0.079305  45.633  < 2e-16 ***
CTA.tot     -0.020068   0.003065  -6.547 1.69e-10 ***
BStotal      0.013469   0.005077   2.653  0.00828 **
---
Signif. codes:  0 '***' 0.001 '**' 0.01 '*' 0.05 '.' 0.1 ' ' 1

Residual standard error: 0.4852 on 426 degrees of freedom
  (57 observations deleted due to missingness)
```

```
Multiple R-squared: 0.1066,    Adjusted R-squared: 0.1024
F-statistic: 25.43 on 2 and 426 DF,  p-value: 3.706e-11
```

From the model summary we can obtain information on model fit (overall F test for significance, R^2 and standard error of the estimate), parameter significance tests, and a summary of residual statistics. As the F test for the overall model is somewhat abbreviated in this output, we can request the entire ANOVA result, including sums of squares and mean squares, by using the anova(Model1.1) function call.

```
Analysis of Variance Table

Response: GPA
            Df  Sum Sq Mean Sq F value     Pr(>F)
CTA.tot      1  10.316 10.3159 43.8125 1.089e-10 ***
BStotal      1   1.657  1.6570  7.0376   0.00828 **
Residuals  426 100.304  0.2355
---
Signif. codes:  0 '***' 0.001 '**' 0.01 '*' 0.05 '.' 0.1 ' ' 1
```

Often in a regression model, we are interested in additional information that the model produces, such as predicted values and residuals. Using the R call attributes() we can get a list of the additional information available for the lm function.

```
attributes(Model1.1)

$names
 [1] "coefficients"  "residuals"     "effects"      "rank"
"fitted.values"
 [6] "assign"        "qr"            "df.residual"  "na.
action"        "xlevels"
[11] "call"          "terms"         "model"

$class
[1] "lm"
```

This is a list of attributes or information that can be pulled out of the fitted regression model. In order to obtain this information from the fitted model, we can call for the particular attribute. For example, if we wished to obtain the predicted GPAs for each individual in the sample, we would simply type the following, followed by the enter key:

```
Model1.1$fitted.values

1        3        4        5        8        9       10       11       12
2.964641 3.125996 3.039668 3.125454 2.852730 3.152391 3.412460 3.011917 2.611103
13       14       15       16       17       19       23       25       26
3.158448 3.298923 3.312121 2.959938 3.205183 2.945928 2.904979 3.226064 3.245318
```

```
       27        28        29        30        31        34        35        37        38
  2.944573  3.171646  2.917635  3.198584  3.206267  3.073204  3.258787  3.118584  2.972594
       39        41        42        43        44        45        46        48        50
  2.870630  3.144980  3.285454  3.386064  2.871713  2.911849  3.166131  3.051511  3.251917
```

Thus, for example, the predicted GPA for subject 1 based on the prediction equation would be 2.96. By the same token, we can obtain the regression residuals with the following command:

```
Model1.1$residuals
```

```
 1                3                4                5                8                9
-0.4646405061  -0.3259956916  -0.7896675749  -0.0254537419   0.4492704297  -0.0283914353
10               11               12               13               14               15
-0.1124596847  -0.5119169570   0.0888967457  -0.6584484215  -0.7989228998  -0.4221207716
16               17               19               23               25               26
-0.5799383942  -0.3051829226  -0.1459275978  -0.8649791080   0.0989363702  -0.2453184879
27               28               29               30               31               34
-0.4445727235   0.7783537067  -0.8176350301   0.1014160133   0.3937331779  -0.1232042042
35               37               38               39               41               42
 0.3412126654   0.4814161689   0.9394056837  -0.6706295541  -0.5449795748  -0.4194540531
43               44               45               46               48               50
-0.4960639410  -0.0717134535  -0.4118490187   0.4338687432   0.7484894275   0.4480825762
```

From this output, we can see that the predicted GPA for the first individual in the sample was approximately 0.465 points above the actual GPA (i.e. observed GPA − predicted GPA = 2.5-2.965).

Interaction Terms in Regression

More complicated regression relationships can also be easily modeled using the lm() function. Let us consider a moderation analysis involving the anxiety measures. In this example, an interaction between cognitive test anxiety and physical anxiety is modeled in addition to the main effects for the two variables. An interaction is simply computed as the product of the interacting variables, so that the moderation model using lm() is defined as:

```
Model1.2 <- lm(GPA ~ CTA.tot + BStotal + CTA.tot*BStotal,
Cassidy)
Model1.2
Call:
lm(formula = GPA ~ CTA.tot + BStotal + CTA.tot * BStotal, data
= Cassidy)

Residuals:
     Min       1Q    Median       3Q      Max
-2.98711  -0.29737   0.01801   0.36340  0.95016
```

```
Coefficients:
                  Estimate Std. Error t value Pr(>|t|)
(Intercept)      3.8977792  0.2307491  16.892  < 2e-16 ***
CTA.tot         -0.0267935  0.0060581  -4.423 1.24e-05 ***
BStotal         -0.0057595  0.0157812  -0.365   0.715
CTA.tot:BStotal  0.0004328  0.0003364   1.287   0.199
---
Signif. codes:  0 '***' 0.001 '**' 0.01 '*' 0.05 '.' 0.1 ' ' 1

Residual standard error: 0.4849 on 425 degrees of freedom
  (57 observations deleted due to missingness)
Multiple R-squared: 0.1101,     Adjusted R-squared: 0.1038
F-statistic: 17.53 on 3 and 425 DF,  p-value: 9.558e-11
```

Here the slope for the interaction is denoted CTA.tot:BStotal, takes the value 0.0004, and is non-significant ($t = 1.287$, $p = 0.199$), which indicates that the level of physical anxiety symptoms (BStotal) does not change or moderate the relationship between cognitive test anxiety (CTA.tot) and GPA.

Categorical Independent Variables

The lm function is also easily capable of incorporating categorical variables into regression. Let us consider an analysis where we predict GPA from cognitive test anxiety (CTA.tot) and the categorical variable, gender. In order for us to incorporate gender into the model, it must be dummy coded such that one category (e.g. male) takes the value of 1, and the other category (e.g. female) takes the value of 0. In this example, we have named the variable Male, where 1 = male and 0 = not male (female). Defining a model using a dummy variable with the lm function then becomes no different from using continuous predictor variables.

```
Model1.3 <- lm(GPA~CTA.tot + Male, Acad)
summary(Model1.3)

Call:
lm(formula = GPA ~ CTA.tot + Male, data = Acad)

Residuals:
     Min       1Q   Median       3Q      Max
-3.01149 -0.29005  0.03038  0.35374  0.96294

Coefficients:
             Estimate Std. Error t value Pr(>|t|)
(Intercept)  3.740318   0.080940  46.211  < 2e-16 ***
CTA.tot     -0.015184   0.002117  -7.173 3.16e-12 ***
```

```
Male            -0.222594    0.047152   -4.721 3.17e-06 ***
---
Signif. codes:  0 '***' 0.001 '**' 0.01 '*' 0.05 '.' 0.1 ' ' 1

Residual standard error: 0.4775 on 437 degrees of freedom
  (46 observations deleted due to missingness)
Multiple R-squared: 0.1364,    Adjusted R-squared: 0.1324
F-statistic: 34.51 on 2 and 437 DF,  p-value: 1.215e-14
```

In this example the slope for the dummy variable Male is negative and significant ($\beta = -0.223$, $p < .001$) indicating that males have a significantly lower mean GPA than do females.

Depending on the format in which the data are stored, the lm function is capable of dummy coding a categorical variable itself. If the variable has been designated as categorical (this often happens if you have read your data from an SPSS file in which the variable is designated as such) then when the variable is used in the lm function it will automatically dummy code the variable for you in your results. For example, if instead of using the Male variable as described above, we used Gender as a categorical variable coded as female and male, we would obtain the following results from the model specification and summary commands.

```
Model1.4 <- lm(GPA~CTA.tot + Gender, Acad)
summary(Model1.4)

Call:
lm(formula = GPA ~ CTA.tot + Gender, data = Acad)

Residuals:
     Min       1Q   Median       3Q      Max
-3.01149 -0.29005  0.03038  0.35374  0.96294

Coefficients:
                Estimate Std. Error t value Pr(>|t|)
(Intercept)     3.740318   0.080940  46.211  < 2e-16 ***
CTA.tot        -0.015184   0.002117  -7.173 3.16e-12 ***
Gender[T.male] -0.222594   0.047152  -4.721 3.17e-06 ***
---
Signif. codes:  0 '***' 0.001 '**' 0.01 '*' 0.05 '.' 0.1 ' ' 1

Residual standard error: 0.4775 on 437 degrees of freedom
  (46 observations deleted due to missingness)
Multiple R-squared: 0.1364,    Adjusted R-squared: 0.1324
F-statistic: 34.51 on 2 and 437 DF,  p-value: 1.215e-14
```

A comparison of results between models Model1.3 and Model1.4 reveals identical coefficient estimates, *p*-values and model fit statistics. The only difference between the two sets of results is that for Model1.4 R has reported the slope as Gender[t.male] indicating that the variable has been automatically dummy coded so that Male is 1 and not Male is 0.

In the same manner, categorical variables consisting of more than two categories can also be easily incorporated into the regression model, either through direct use of the categorical variable or dummy coded prior to analysis. In the following example, the variable Ethnicity includes three possible groups, African American, Other, and Caucasian. By including this variable in the model call, we are implicitly requesting that R automatically dummy code it for us.

```
GPAmodel1.5 <- lm(GPA~CTA.tot + Ethnicity, Acad)
summary(GPAmodel1.5)

Call:
lm(formula = GPA ~ CTA.tot + Ethnicity, data = Acad)

Residuals:
    Min       1Q    Median       3Q       Max
-2.95019 -0.30021   0.01845  0.37825   1.00682

Coefficients:
                                Estimate Std. Error t value
Pr(>|t|)
(Intercept)                     3.670308   0.079101  46.400   <
2e-16 ***
CTA.tot                        -0.015002   0.002147  -6.989
1.04e-11 ***
Ethnicity[T.African American]  -0.482377   0.131589  -3.666
0.000277 ***
Ethnicity[T.Other]             -0.151748   0.136150  -1.115
0.265652
---
Signif. codes:  0 '***' 0.001 '**' 0.01 '*' 0.05 '.' 0.1 ' ' 1

Residual standard error: 0.4821 on 436 degrees of freedom
  (46 observations deleted due to missingness)
Multiple R-squared: 0.1215,      Adjusted R-squared: 0.1155
F-statistic: 20.11 on 3 and 436 DF,  p-value: 3.182e-12
```

Given that we have slopes for African American and Other, we know that Caucasian serves as the reference category, which is coded as 0. Results indicate a significant negative slope for African American ($\beta = -.482$, $p < .001$), and a non-significant slope for Other ($\beta = -.152$, $p > .05$) indicating that African Americans have a significantly lower GPA than Caucasians, but the Other ethnicity category was not significantly different from Caucasian in terms of GPA.

Finally, let us consider some issues associated with allowing R to auto dummy code categorical variables. First, R will always auto dummy code the first category listed as the reference category. If a more theoretically suitable dummy coding scheme is desired, it will be necessary to either order the categories so that the desired reference category is first, or simply recode into dummy variables manually. Also, it is important to keep in mind that auto dummy coding only occurs when the variable is labeled in the system as categorical. This will occur automatically if the categories themselves are coded as letters. However, if a categorical variable is coded 1, 2 or 1, 2, 3 but not specifically designated as categorical, the system will view it as continuous and treat it as such. In order to ensure that a variable is treated as categorical when that is what we desire, we simply use the as.factor command. For the Male variable, in which males are coded as 1 and females as 0, we would type the following:

```
Male<-as.factor(Male)
```

We would then be able to assume the variable Male as categorical. In addition, if the dummy variable has only two levels, as is the case with Male, then it need not be converted to a categorical factor, as the results obtained when using it in the regression analysis will be identical either way.

Checking Regression Assumptions with R

When checking assumptions for linear regression models, it is often desirable to create a plot of the residuals. Diagnostic residual plots can be easily obtained through use of the residualPlots function from the car R package, which we would need to install in our R workspace as explained in the chapter introducing working with R. Let us again return to Model1.1 predicting GPA from cognitive test anxiety and physical anxiety symptoms. Once the regression model is created (Model1.1), we can easily obtain diagnostic residuals scatterplots using the following command:

```
Library(car)
residualPlots(Model1.1)
```

This command will produce scatterplots of the Pearson residuals against each predictor variable as well as against the fitted values. In addition, the residualPlots command will provide lack of fit tests in which a *t*-test for the predictor squared is computed and a fit line added to the plot to help check for nonlinear patterns in the data. A Tukey's test for nonadditivity is also computed for the plot of residuals against the fitted values for further information regarding adequacy of model fit, along with a lack of fit test for each predictor. Tukey's statistic is obtained by adding the squares of the fitted values to the original regression model. It tests the null hypothesis that the model is additive and that no interactions exist among the independent variables (Tukey,

1949). A non-significant result, such as that found for this example, indicates that no interaction is required in the model. The other tests included here are for the squared term of each independent variable. For example, given that the Test stat for CTA.tot or BStotal are not significant, we can conclude that neither of these variables has a quadratic relationship with GPA (Figure 1.1).

```
residualPlots (Model1.1)
```

	Test stat	Pr(>|t|)
CTA.tot	0.607	0.544
BStotal	0.762	0.447
Tukey test	0.301	0.764

The residualPlots command provides plots with the residuals on the y-axis of the graphs, and the values of each independent variable respectively, on the x-axis for the first two graphs, and the fitted values on x for the last graph. In addition, curves were fit linking the x and y-axes for each graph. The researcher would examine these graphs in order to assess two

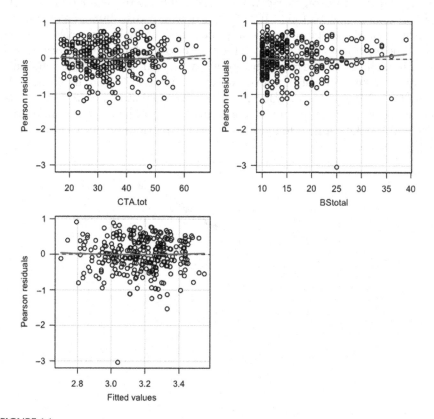

FIGURE 1.1
Diagnostic residual plots for regression model predicting GPA from CTA.tot and Bstotal.

assumptions about the data. First, the assumption of homogeneity of variance can be checked through an examination of the residual by a fitted plot.
If the assumption holds, this plot should display a formless cloud of data
points with no discernible shape, which are equally spaced across all values
of x. In addition, the linearity of the relationships between each independent variable and the dependent is assessed through an examination of the
plots involving them. For example, it is appropriate to assume linearity for
BStotal if the residual plots show no discernible pattern, which can be further understood with an examination of the fitted line. If this line is essentially flat, as is the case here, then we can conclude that any relationship that
is present between BStotal and GPA is only linear.

In addition to linearity and homogeneity of variance, it is also important to
determine whether the residuals follow a normal distribution, as is assumed
in regression analysis. In order to check the normality of residuals assumption, Q-Q plots (quantile-quantile plots) are typically used. The qqPlot
function from the car package may be used to easily create Q-Q plots of
regression models once run. The interpretation of the Q-Q plot is quite simple. Essentially, the graph displays the data as they actually are on the x-axis
and the data as they would be if they were normally distributed on the y-axis.
The individual data points are represented in R by the black circles, and the
solid line represents the data when they conform perfectly to the normal distribution. Therefore, the closer the observed data (circles) are to the solid line,
the more closely the data conform to the normal distribution. In addition, R
provides a 95% confidence interval for the line, so that when the data points
fall within it, they are deemed to conform to the normal distribution. In this
example, the data appear to follow the normal distribution fairly closely.
qqPlot(Model1.1)

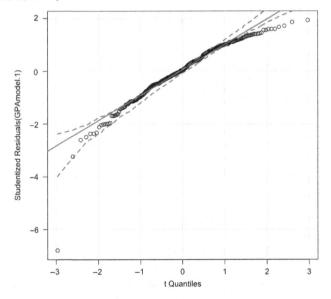

Summary

Chapter 1 introduced the reader to the basics of linear modeling using R. This treatment was purposely limited, as there are a number of good texts available on this subject, and it is not the main focus of this book. However, many of the core concepts presented here for the GLM apply to multilevel modeling as well, and thus are of key importance as we move into these more complex analyses. In addition, much of the syntactical framework presented here will reappear in subsequent chapters. In particular, readers should leave this chapter comfortable with interpretation of coefficients in linear models, as well as the concept of variance explained in an outcome variable. We would encourage you to return to this chapter frequently as needed in order to reinforce these basic concepts. In addition, we would recommend that you also refer to the initial chapter dealing with the basics of using R when questions regarding data management and installation of specific R libraries become an issue. Next, in Chapter 2 we will turn our attention to the conceptual underpinnings of multilevel modeling before delving into their estimation in Chapters 3 and 4.

2

An Introduction to Multilevel Data Structure

Nested Data and Cluster Sampling Designs

In Chapter 1, we considered the standard linear model that underlies such common statistical methods as regression and analysis of variance (ANOVA; i.e. the general linear model). As we noted, this model rests on several primary assumptions regarding the nature of the data in the population. Of particular importance in the context of multilevel modeling is the assumption of independently distributed error terms for the individual observations within the sample. This assumption essentially means that there are no relationships among individuals in the sample for the dependent variable, *once the independent variables in the analysis are accounted for.* In the example we described in Chapter 1, this assumption was indeed met, as the individuals in the sample were selected randomly from the general population. Therefore, there was nothing linking their dependent variable values other than the independent variables included in the linear model. However, in many cases the method used for selecting the sample does create the correlated responses among individuals. For example, a researcher interested in the impact of a new teaching method on student achievement might randomly select schools for placement in either a treatment or control group. If school A is placed into the treatment condition, all students within the school will also be in the treatment condition – this is a cluster randomized design, in that the clusters and not the individuals are assigned to a specific group. Furthermore, it would be reasonable to assume that the school itself, above and beyond the treatment condition, would have an impact on the performance of the students. This impact would manifest itself as correlations in achievement test scores among individuals attending that school. Thus, if we were to use a simple one-way ANOVA to compare the achievement test means for the treatment and control groups with such cluster sampled data, we would likely be violating the assumption of independent errors because a factor beyond treatment condition (in this case the school) would have an additional impact on the outcome variable.

We typically refer to the data structure described above as nested, meaning that individual data points at one level (e.g. student) appear in only one level of a higher-level variable such as school. Thus, students are nested within school. Such designs can be contrasted with a crossed data structure whereby individuals at the first level appear in multiple levels of the second variable. In our example, students might be crossed with after-school organizations if they are allowed to participate in more than one. For example, a given student might be on the basketball team as well as in the band. The focus of this book is almost exclusively on nested designs, which give rise to multilevel data. Other examples of nested designs might include a survey of job satisfaction for employees from multiple depart-ments within a large business organization. In this case, each employee works within only a single division in the company, which leads to a nested design. Furthermore, it seems reasonable to assume that employ-ees working within the same division will have correlated responses on the satisfaction survey, as much of their view regarding the job would be based exclusively upon experiences within their division. For a third such example, consider the situation in which clients of several psycho-therapists working in a clinic are asked to rate the quality of each of their therapy sessions. In this instance, there exist three levels in the data: time, in the form of individual therapy session, client, and therapist. Thus, ses-sion is nested in client, who in turn is nested within therapist. All of this data structure would be expected to lead to correlated scores on a therapy-rating instrument.

Intraclass Correlation

In cases where individuals are clustered or nested within a higher-level unit (e.g. classrooms, schools, school districts), it is possible to estimate the cor-relation among individuals' scores within the cluster/nested structure using the intraclass correlation (denoted ρ_I in the population). The ρ_I is a measure of the proportion of variation in the outcome variable that occurs between groups versus the total variation present and ranges from 0 (no variance between clusters) to 1 (variance between clusters but not within cluster vari-ance). ρ_I can also be conceptualized as the correlation for the dependent mea-sure for two individuals randomly selected from the same cluster. It can be expressed as

$$\rho_I = \frac{\tau^2}{\tau^2 + \sigma^2} \tag{2.1}$$

where
τ^2 = Population variance between clusters
σ^2 = Population variance within clusters

Higher values of ρ_I indicate that a greater share of the total variation in the outcome measure is associated with cluster membership; i.e. there is a relatively strong relationship among the scores for two individuals from the same cluster. Another way to frame this issue is that individuals within the same cluster (e.g. school) are more alike on the measured variable than they are like those in other clusters.

It is possible to estimate τ^2 and σ^2 using sample data, and thus it is also possible to estimate ρ_I. Those familiar with ANOVA will recognize these estimates as being related (though not identical) to the sum of squared terms. The sample estimate for variation within clusters is simply

$$\hat{\sigma}^2 = \frac{\sum_{j=1}^{C}(n_j - 1)S_j^2}{N - C}$$

where

$$S_j^2 = \text{variance within cluster } j = \frac{\sum_{i=1}^{n_j}(y_{ij} - \bar{y}_j)}{(n_j - 1)}$$

n_j = sample size for cluster j
N = total sample size
C = total number of clusters

$$\hat{\sigma}^2 = \frac{\sum_{j=1}^{C}(n_j - 1)S_j^2}{N - C} \tag{2.2}$$

where

$$S_j^2 = \frac{\sum_{j=1}^{n_j}(y_{ij} - \bar{y}_j)^2}{(n_j - 1)}$$

n_j = sample size for cluster j
N = total sample size
C = total number of clusters

In other words, σ^2 is simply the weighted average of within cluster variances.

Estimation of τ^2 involves a few more steps, but it is not much more complex than what we have seen for σ^2. In order to obtain the sample estimate for variation between clusters, $\hat{\tau}^2$, we must first calculate the weighted between-cluster variance.

$$\hat{S}_B^2 = \frac{\displaystyle\sum_{j=1}^{C} n_j (\bar{y}_j - \bar{y})^2}{\tilde{n}(C-1)} \tag{2.3}$$

where

$\bar{y}_j$ = mean on response variable for cluster j
$\bar{y}$ = overall mean on response variable

$$\tilde{n} = \frac{1}{C-1}\left[N - \frac{\displaystyle\sum_{j=1}^{C} n_j^2}{N} \right]$$

We cannot use S_B^2 as a direct estimate of τ^2 because it is impacted by the random variation among subjects within the same clusters. Therefore, in order to remove this random fluctuation, we will estimate the population between-cluster variance as

$$\hat{\tau}^2 = S_B^2 - \frac{\hat{\sigma}^2}{\tilde{n}}. \tag{2.4}$$

Using these variance estimates, we can in turn calculate the sample estimate of ρ_I:

$$\hat{\rho}_I = \frac{\hat{\tau}^2}{\hat{\tau}^2 + \hat{\sigma}^2}. \tag{2.5}$$

Note that Equation (2.5) assumes that the clusters are of equal size. Clearly, such will not always be the case, in which case this equation will not hold. However, the purpose for its inclusion here is to demonstrate the principle underlying the estimation of ρ_I, which holds even as the equation might change.

In order to illustrate estimation of ρ_I, let us consider the following dataset. Achievement test data were collected from 10,903 third-grade examinees nested within 160 schools. School sizes range from 11 to 143, with a mean size of 68.14. In this case, we will focus on the reading achievement test score, and will use data from only five of the schools, in order to make the calculations by hand easy to follow. First, we will estimate $\hat{\sigma}^2$. To do so, we must estimate the variance in scores within each school. These values appear in Table 2.1.

TABLE 2.1

School Size, Mean, and Variance of Reading Achievement Test

School	N	Mean	Variance
767	58	3.952	5.298
785	29	3.331	1.524
789	64	4.363	2.957
815	39	4.500	6.088
981	88	4.236	3.362
Total	278	4.149	3.916

Using these variances and sample sizes, we can calculate $\hat{\sigma}^2$ as

$$\hat{\sigma}^2 = \frac{\sum_{j=1}^{C}(n_j-1)S_j^2}{N-C} = \frac{(58-1)5.3+(29-1)1.5+(64-1)2.9+(39-1)6.1+(88-1)3.4}{278-5}$$

$$= \frac{302.1+42+182.7+231.8+295.8}{273} = \frac{1054.4}{273} = 3.9$$

The school means, which are needed in order to calculate S_B^2, appear in Table 2.2 as well. First, we must calculate n:

$$\tilde{n} = \frac{1}{C-1}\left(N-\frac{\sum_{j=1}^{C}n_j^2}{N}\right) = \frac{1}{5-1}\left(278-\frac{58^2+29^2+64^2+39^2+88^2}{278}\right) = \frac{1}{4}(278-63.2)$$

$$= 53.7$$

TABLE 2.2

Between Subjects Intercept and Slope, and within Subjects Variation on These Parameters by School

School	Intercept	U_{0j}	Slope	U_{1j}
1	1.230	−1.129	0.552	0.177
2	2.673	0.314	0.199	−0.176
3	2.707	0.348	0.376	0.001
4	2.867	0.508	0.336	−0.039
5	2.319	−0.040	0.411	0.036
Overall	2.359		0.375	

Using this value, we can then calculate S_B^2 for the five schools in our small sample using Equation (2.3):

$$\frac{58(3.952-4.149)^2+29(3.331-4.149)^2+64(4.363-4.149)^2+39(4.500-4.149)^2+88(4.236-4.149)^2}{53.7(5-1)}$$

$$=\frac{2.251+19.405+2.931+4.805+0.666}{214.8}=\frac{30.057}{214.800}=0.140$$

We can now estimate the population between-cluster variance, τ^2, using Equation (2.4):

$$0.140-\frac{3.9}{53.7}=0.140-0.073=0.067$$

We have now calculated all of the parts that we need to estimate ρ_I for the population,

$$\hat{\rho}_I=\frac{0.067}{0.067+3.9}=0.017$$

This result indicates that there is very little correlation of examinees' test scores within the schools. We can also interpret this value as the proportion of variation in the test scores that is accounted for by the schools.

Given that $\hat{\rho}_I$ is a sample estimate, we know that it is subject to sampling variation, which can be estimated with a standard error as in Equation (2.6):

$$s_{\rho_I}=(1-\rho_I)(1+(n-1)\rho_I)\sqrt{\frac{2}{\sqrt{n(n-1)(N-1)}}}. \qquad (2.6)$$

The terms in 2.6 are as defined previously, and the assumption is that all clusters are of equal size. As noted earlier in the chapter, this latter condition is not a requirement, however, and an alternative formulation exists for cases in which it does not hold. However, 2.6 provides sufficient insight for our purposes into the estimation of the standard error of the ICC.

The ICC is an important tool in multilevel modeling, in large part because it is an indicator of the degree to which the multilevel data structure might impact the outcome variable of interest. Larger values of the ICC are indicative of a greater impact of clustering. Thus, as the ICC increases in value, we must be more cognizant of employing multilevel modeling strategies in our data analysis. In the next section, we will discuss the problems associated with ignoring this multilevel structure, before we turn our attention to methods for dealing with it directly.

Pitfalls of Ignoring Multilevel Data Structure

When researchers apply standard statistical methods to multilevel data, such as the regression model described in Chapter 1, the assumption of independent errors is violated. For example, if we have achievement test scores from a sample of students who attend several different schools, it would be reasonable to believe that those attending the same school will have scores that are more highly correlated with one another than they are with scores from students attending other schools. This within-school correlation would be due, for example, to having a common set of teachers, a common teaching curriculum, a single set of administrative policies, and a common community, among numerous other reasons. The within-school correlation will in turn result in an inappropriate estimate of the standard errors for the model parameters, which in turn will lead to errors of statistical inference, such as *p*-values smaller than they really should be and the resulting rejection of null effects above the stated Type I error rate, regarding the parameters. Recalling our discussion in Chapter 1, the test statistic for the null hypothesis of no relationship between the independent and dependent variable is simply the regression coefficient divided by the standard error. If the standard error is underestimated, this will lead to an overestimation of the test statistic, and therefore statistical significance for the parameter in cases where it should not be; i.e. a Type I error at a higher rate than specified. Indeed, the underestimation of the standard error will occur unless τ^2 is equal to 0.

In addition to the underestimation of the standard error, another problem with ignoring the multilevel structure of data is that we may miss important relationships involving each level in the data. Recall that in our example, there are two levels of sampling: students (level 1) are nested in schools (level 2). Specifically, by *not* including information about the school, for example, we may well miss important variables at the school level that help to explain performance at the examinee level. Therefore, beyond the known problem with misestimating standard errors, we also proffer an incorrect model for understanding the outcome variable of interest. In the context of MLMs, inclusion of variables at each level is relatively simple, as are interactions among variables at different levels. This greater model complexity in turn may lead to greater understanding of the phenomenon under study.

Multilevel Linear Models

In the following section we will review some of the core ideas that underlie multilevel linear models (MLM). Our goal is to familiarize the reader with terms that will repeat themselves throughout the book, and to do so in a relatively nontechnical fashion. We will first focus on the difference between random and fixed effects, after which we will discuss the basics of parameter estimation, focusing on the two most commonly used methods, maximum likelihood and restricted maximum likelihood, and conclude with a review of assumptions underlying MLMs, and an overview of how they are most frequently used, with examples. In this section, we will also address the issue of centering, and explain why it is an important concept in MLM. After reading the rest of this chapter, the reader will have sufficient technical background on MLMs to begin using the R software package for fitting MLMs of various types.

Random Intercept

As we transition from the one-level regression framework of Chapter 1 to the MLM context, let's first revisit the basic simple linear regression model of Equation (1.1), $y = \beta_0 + \beta_1 x + \varepsilon$. Here, the dependent variable y is expressed as a function of an independent variable, x, multiplied by a slope coefficient, β_1, an intercept, β_0, and random variation from subject to subject, ε. We defined the intercept as the conditional mean of y when the value of x is 0. In the context of a single-level regression model such as this, there is one intercept that is common to all individuals in the population of interest. However, when individuals are clustered together in some fashion (e.g. within classrooms, schools, organizational units within a company), there will potentially be a separate intercept for each of these clusters; that is, there may be different means for the dependent variable for $x = 0$ across the different clusters. We say *potentially* here because if there is in fact no cluster effect, then the single intercept model of 1.1 will suffice. In practice, assessing whether there are different means across the clusters is an empirical question, which we describe below. It should also be noted that in this discussion we are considering only the case where the intercept is cluster specific, but it is also possible for β_1 to vary by group, or even other coefficients from more complicated models.

Allowing for group-specific intercepts and slopes leads to the following notation commonly used for the level 1 (micro level) model in multilevel modeling:

$$y_{ij} = \beta_{0j} + \beta_{1j}x + \varepsilon_{ij} \tag{2.7}$$

where the subscripts ij refer to the ith individual in the jth cluster. As we continue our discussion of multilevel modeling notation and structure, we will begin with the most basic multilevel model: predicting the outcome from just an intercept which we will allow to vary randomly for each group.

$$y_{ij} = \beta_{0j} + \varepsilon_{ij}. \tag{2.8}$$

Allowing the intercept to differ across clusters, as in Equation (2.8), leads to the random intercept which we express as

$$\beta_{0j} = \gamma_{00} + U_{0j}. \tag{2.9}$$

In this framework, γ_{00} represents an average or general intercept value that holds across clusters, whereas U_{0j} is a group-specific effect on the intercept. We can think of γ_{00} as a fixed effect because it remains constant across all clusters, and U_{0j} is a random effect because it varies from cluster to cluster. Therefore, for an MLM we are interested not only in some general mean value for y when x is 0 for all individuals in the population (γ_{00}), but also in the deviation between the overall mean and the cluster-specific effects for the intercept (U_{0j}). If we go on to assume that the clusters are a random sample from the population of all such clusters, then we can treat the U_{0j} as a kind of residual effect on y_{ij}, very similar to how we think of ε. In that case, U_{0j} is assumed to be drawn randomly from a population with a mean of 0 (recall U_{0j} is a deviation from the fixed effect) and a variance, τ^2. Furthermore, we assume that τ^2 and σ^2, the variance of ε, are uncorrelated. We have already discussed τ^2 and its role in calculating $\hat{\rho}_I$. In addition, τ^2 can also be viewed as the impact of the cluster on the dependent variable, and therefore testing it for statistical significance is equivalent to testing the null hypothesis that cluster (e.g. school) has no impact on the dependent variable. If we substitute the two components of the random intercept into the regression model, we get

$$y = \gamma_{00} + U_{0j} + \beta_1 x + \varepsilon. \tag{2.10}$$

Equation (2.10) is termed the full or composite model in which the multiple levels are combined into a unified equation.

Often in MLM, we begin our analysis of a dataset with this simple random intercept model, known as the null model, which takes the form

$$y_{ij} = \gamma_{00} + U_{0j} + \varepsilon_{ij}. \tag{2.11}$$

While the null model does not provide information regarding the impact of specific independent variables on the dependent, it does yield important information regarding how variation in y is partitioned between variance among the individuals σ^2 and variance among the clusters τ^2. The total variance of y is simply the sum of σ^2 and τ^2. In addition, as we have already seen, these values can be used to estimate ρ_I. The null model, as will be seen in later sections, is also used as a baseline for model building and comparison.

Random Slopes

It is a simple matter to expand the random intercept model in 2.9 to accommodate one or more independent predictor variables. As an example, if we add a single predictor (x_{ij}) at the individual level (level 1) to the model, we obtain

$$y_{ij} = \gamma_{00} + \gamma_{10}x_{ij} + U_{0j} + \varepsilon_{ij}. \qquad (2.12)$$

This model can also be expressed in two separate levels as:
   Level 1:

$$y_{ij} = \beta_{0j} + \beta_{1j}x + \varepsilon_{ij} \qquad (2.13)$$

   Level 2:

$$\beta_{0j} = \gamma_{00} + U_{0j} \qquad (2.14)$$

$$\beta_{1j} = \gamma_{10} \qquad (2.15)$$

This model now includes the predictor and the slope relating it to the dependent variable, γ_{10}, which we acknowledge as being at level 1 by the subscript 10. We interpret γ_{10} in the same way that we did β_1 in the linear regression model; i.e. a measure of the impact on y of a 1-unit change in x. In addition, we can estimate ρ_I exactly as before, though now it reflects the correlation between individuals from the same cluster after controlling for the independent variable, x. In this model, both γ_{10} and γ_{00} are fixed effects, while σ^2 and τ^2 remain random.

One implication of the model in 2.12 is that the dependent variable is impacted by variation among individuals (σ^2), variation among clusters (τ^2), an overall mean common to all clusters (γ_{00}), and the impact of the independent variable as measured by γ_{10}, which is also common to all clusters. In practice there is no reason that the impact of x on y would need to be common for all clusters, however. In other words, it is entirely possible that rather than a single γ_{10} common to all clusters, there is actually a unique effect for the cluster of $\gamma_{10} + U_{1j}$, where γ_{10} is the average relationship of x with y across clusters, and U_{1j} is the cluster-specific variation of the relationship between the two variables. This cluster -specific effect is assumed to have a mean of 0 and to vary randomly around γ_{10}. The random slopes model is

$$y_{ij} = \gamma_{00} + \gamma_{10}x_{ij} + U_{0j} + U_{1j}x_{ij} + \varepsilon_{ij}. \qquad (2.16)$$

Written in this way, we have separated the model into its fixed $(\gamma_{00} + \gamma_{10}x_{ij})$ and random $(U_{0j} + U_{1j}x_{ij} + \varepsilon_{ij})$ components. Model 2.16 simply states that there is an interaction between cluster and x, such that the relationship of x and y is not constant across clusters.

Heretofore we have discussed only one source of between-group variation, which we have expressed as τ^2, and which is the variation among clusters in the intercept. However, Model2.16 adds a second such source of between-group variance in the form of U_{1j}, which is cluster variation on the slope relating the independent and dependent variables. In order to differentiate between these two sources of between-group variance, we now denote the variance of U_{0j} as τ_0^2 and the variance of U_{1j} as τ_1^2. Furthermore, within clusters, we expect U_{1j} and U_{0j} to have a covariance of τ_{01}. However, across different clusters, these terms should be independent of one another, and in all cases it is assumed that ε remains independent of all other model terms. In practice, if we find that τ_1^2 is not 0, we must be careful in describing the relationship between the independent and dependent variables, as it is not the same for all clusters. We will revisit this idea in subsequent chapters. For the moment, however, it is most important to recognize that variation in the dependent variable, y, can be explained by several sources, some fixed and others random. In practice, we will most likely be interested in estimating all of these sources of variability in a single model.

As a means for further understanding the MLM, let's consider a simple example using the five schools described above. In this context, we are interested in treating a reading achievement test score as the dependent variable and a vocabulary achievement test score as the independent variable. Remember that students are nested within schools so that a simple regression analysis will not be appropriate. In order to understand what is being estimated in the context of MLM, we can obtain separate intercept and slope estimates for each school, which appear in Table 2.2.

Given that the schools are of the same sample size, the estimate of γ_{00}, the average intercept value is 2.359, and the estimate of the average slope value, γ_{10}, is 0.375. Notice that for both parameters, the school values deviate from these means. For example, the intercept for school 1 is 1.230. The difference between this value and 2.359, -1.129, is U_{0j} for that school. Similarly, the difference between the average slope value of 0.375 and the slope for school 1, 0.552, is 0.177, which is U_{1j} for this school. Table 2.2 includes U_{0j} and U_{1j} values for each of the schools. The differences in slopes also provide information regarding the relationship between vocabulary and reading test scores. For all of the schools this relationship was positive, meaning that students who scored higher on vocabulary also scored higher on reading. However, the strength of this relationship was weaker for school 2 than for school 1, as an example.

Given the values in Table 2.2, it is also possible to estimate the variances associated with U_{1j} and U_{0j}, τ_1^2 and τ_0^2, respectively. Again, because the schools in this example had the same number of students, the calculation of these variances is a straightforward matter, using

$$\frac{\sum \left(U_{1j} - \overline{U}_{1.}\right)^2}{J-1} \tag{2.17}$$

for the slopes, and an analogous equation for the intercept random variance. Doing so, we obtain $\tau_0^2 = 0.439$, and $\tau_1^2 = 0.016$. In other words, much more of the variance in the dependent variable is accounted for by variation in the intercepts at the school level than is accounted for by variation in the slopes. Another way to think of this result is that the schools exhibited greater differences among one another in the mean level of achievement as compared to differences in the impact of x on y.

The actual practice of obtaining these variance estimates using the R environment for statistical computing and graphics and interpreting their meaning are subjects for the coming chapters. Before discussing the practical nuts and bolts of conducting this analysis, we will first examine the basics for how parameters are estimated in the MLM framework using maximum likelihood and restricted maximum likelihood algorithms. While similar in spirit to the simple calculations demonstrated above, they are different in practice and will yield somewhat different results than those we would obtain using least squares, as above. Prior to this discussion, however, there is one more issue that warrants our attention as we consider the practice of MLM, namely variable centering.

Centering

Centering simply refers to the practice of subtracting the mean of a variable from each individual value. This implies the mean for the sample of the centered variables is 0, and implies that each individual's (centered) score represents a deviation from the mean, rather than whatever meaning its raw value might have. In the context of regression, centering is commonly used, for example, to reduce collinearity caused by including an interaction term in a regression model. If the raw scores of the independent variables are used to calculate the interaction, and then both the main effects and interaction terms are included in the subsequent analysis, it is very likely that collinearity will cause problems in the standard errors of the model parameters. Centering is a way to help avoid such problems (e.g. Iversen, 1991). Such issues are also important to consider in MLMs, in which interactions are frequently employed. In addition, centering is also a useful tool for avoiding collinearity caused by highly correlated random intercepts and slopes in MLMs (Wooldridge, 2004). Finally, centering provides a potential advantage in terms of interpretation of results. Remember from our discussion in Chapter 1 that the intercept is the value of the dependent variable when the independent variable is set equal to 0. In many applications the independent variable cannot reasonably be 0 (e.g. a measure of vocabulary), however, which essentially renders the intercept as a necessary value for fitting the regression line but not one that has a readily interpretable value. However, when x has been centered, the intercept takes on the value of the dependent variable when the independent is at its mean. This is a much more useful interpretation for researchers in many situations, and yet another reason why centering is an important aspect of modeling, particularly in the multilevel context.

Probably the most common approach to centering is to calculate the difference between each individual's score and the overall, or grand mean across the entire sample. This *grand mean centering* is certainly the most commonly used in practice (Bickel, 2007). It is not, however, the only manner in which data can be centered. An alternative approach, known as *group mean centering*, is to calculate the difference between each individual score and the mean of the cluster to which they belong. In our school example, grand mean centering would involve calculating the difference between each score and the overall mean across schools, while group mean centering would lead the researcher to calculate the difference between each score and the mean for their school. While there is some disagreement in the literature regarding which approach might be best at reducing the harmful effects of collinearity (Bryk and Raudenbush, 2002; Snijders and Bosker, 1999), researchers have demonstrated that in most cases either will work well in this regard (Kreft, de Leeuw, and Aiken, 1995). Therefore, the choice of which approach to use must be made on substantive grounds regarding the nature of the relationship between x and y. By using grand mean centering, we are implicitly comparing individuals to one another (in the form of the overall mean) across the entire sample. On the other hand, when using group mean centering, we are placing each individual in their relative position on x within their cluster. Thus, in our school example, using the group mean centered values of vocabulary in the analysis would mean that we are investigating the relationship between one's relative vocabulary score in their school and their reading score. In contrast, the use of grand mean centering would examine the relationship between one's relative standing in the sample as a whole on vocabulary and the reading score. This latter interpretation would be equivalent conceptually (though not mathematically) to using the raw score, while the group mean centering would not. Throughout the rest of this book, we will use grand mean centering by default, per recommendations by Hox (2002), among others. At times, however, we will also demonstrate the use of group mean centering in order to illustrate how it provides different results, and for applications in which interpretation of the impact of an individual's relative standing in their cluster might be more useful than their relative standing in the sample as a whole.

Basics of Parameter Estimation with MLMs

Heretofore, when we have discussed estimation of model parameters, it has been in the context of least squares, which serve as the underpinnings of OLS and related linear models. However, as we move from these fairly simple applications to more complex models, OLS is not typically the optimal approach to use for parameter estimation. Instead, we will rely on maximum-likelihood estimation (MLE) and restricted maximum likelihood (REML). In the following sections we review these approaches to estimation from a conceptual basis, focusing on the generalities of how they work, what they assume about the data, and how they differ from one another. For the technical details we refer the interested reader to Bryk and Raudenbush (2002) or de Leeuw and Meijer (2008), both of which provide excellent resources for those desiring a more in-depth coverage of these methods. Our purpose here is to provide the reader with a conceptual understanding that will aid in their understanding of application of MLMs in practice.

Maximum Likelihood Estimation

MLE has as its primary goal the estimation of population model parameters that maximize the likelihood of our obtaining the sample that we in fact obtained. In other words, the estimated parameter values should maximize the likelihood of our particular sample. From a practical perspective, identifying such sample values takes place through the comparison of the observed data with that predicted by the model associated with the parameter values. The closer the observed and predicted values are to one another, the greater the likelihood that the observed data arose from a population with parameters close to those used to generate the predicted values. In practice, MLE is an iterative methodology in which the algorithm searches for those parameter values that will maximize the likelihood of the observed data (i.e. produce predicted values that are as close as possible to the observed), and as such can be computationally intensive, particularly for complex models and large samples.

Restricted Maximum Likelihood Estimation

There exists a variant of MLE, restricted maximum likelihood, that has been shown to be more accurate with regard to the estimation of variance parameters than is MLE (Kreft and De Leeuw, 1998). In particular, the two methods differ with respect to how degrees of freedom are calculated in the estimation of variances. As a simple example, the sample variance is typically calculated by dividing the sum of squared differences between individual values and the mean by the number of observations minus 1, so as to have an unbiased estimate. This is a REML estimate of variance. In contrast, the MLE variance is calculated by dividing the sum of squared differences by the total sample size, leading to a smaller variance estimate than REML and, in fact, one that is biased in finite samples. In the context of multilevel modeling, REML takes into account the number of parameters being estimated in the model when determining the appropriate degrees of freedom for the estimation of the random components such as the parameter variances described above. In contrast, MLE does not account for these, leading to an underestimate of the variances that does not occur with REML. For this reason, REML is generally the preferred method for estimating multilevel models, though for testing variance parameters (or any random effect) it is necessary to use MLE (Snijders and Bosker, 1999). It should be noted that as the number of level-2 clusters increases, the difference in value for MLE and REML estimates becomes very small (Snijders and Bosker, 1999).

Assumptions Underlying MLMs

As with any statistical model, the appropriate use of MLMs requires that several assumptions about the data hold true. If these assumptions are not met, the model parameter estimates may not be trustworthy, just as would be the case with standard linear regression, which was reviewed in Chapter 1. Indeed, while they differ somewhat from the assumptions for the single-level models, the assumptions underlying MLMs are akin to those for the simpler models. In this section, we provide an introduction to these assumptions and their implications for researchers using MLMs, and in subsequent chapters we describe methods for checking the validity of these assumptions for a given set of data.

First, we assume that the level-2 residuals are independent between clusters. In other words, there is an assumption that the random intercept and slope(s) at level 2 are independent of one another across clusters. Second, the level 2 intercepts and coefficients are assumed to be independent of the level-1 residuals; i.e. the errors for the cluster-level estimates are unrelated to errors at the individual level. Third, the level-1 residuals are normally distributed and have a constant variance. This assumption is very similar to the one we make about residuals in the standard linear regression model. Fourth, the level-2 intercept and slope(s) have a multivariate normal distribution with a constant covariance matrix. Each of these assumptions can be directly assessed for a sample, as we shall see in forthcoming chapters. Indeed, the methods for checking the MLM assumptions are not very different than those for checking the regression model that we used in Chapter 1.

Overview of Two-Level MLMs

To this point, we have described the specific terms that make up the MLM, including the level-1 and level-2 random effects and residuals. We will close out this chapter introducing the MLM by considering an example of two- and three-level MLMs, and the use of MLMs with longitudinal data. This discussion should prepare the reader for subsequent chapters in which we consider the application of R to the estimation of specific MLMs. First, let us consider the two-level MLM, parts of which we have already described previously in the chapter.

Previously, in Equation (2.16), we considered the random slopes model $y_{ij} = \gamma_{00} + \gamma_{10}x_{ij} + U_{0j} + U_{1j}x_{ij} + \varepsilon_{ij}$ in which the dependent variable, y_{ij} (reading achievement), was a function of an independent variable x_{ij} (vocabulary test score), as well as random error at both the examinee and school level. We can extend this model a bit further by including multiple independent variables at both level 1 (examinee) and level 2 (school). Thus, for example, in addition to

ascertaining the relationship between an individual's vocabulary and reading scores, we can also determine the degree to which the average vocabulary score at the school as a whole is related to an individual's reading score.
This model would essentially have two parts, one explaining the relationship between the individual level vocabulary (x_{ij}) and reading, and the other
explaining the coefficients at level 1 as a function of the level-2 predictor, average vocabulary score (z_j). The two parts of this model are expressed as:
Level 1:

$$y_{ij} = \beta_{0j} + \beta_{1j}x_{ij} + \varepsilon_{ij} \tag{2.18}$$

and

Level 2:

$$\beta_{hj} = \gamma_{h0} + \gamma_{h1}z_j + U_{hj}. \tag{2.19}$$

The additional piece of the equation in 2.19 is $\gamma_{h1}z_j$, which represents the slope
for (γ_{h1}), and value of the average vocabulary score for the school (z_j). In other
words, the mean school performance is related directly to the coefficient
linking the individual vocabulary score to the individual reading score. For
our specific example, we can combine 2.18 and 2.19 in order to obtain a single
equation for the two-level MLM.

$$y_{ij} = \gamma_{00} + \gamma_{10}x_{ij} + \gamma_{01}Z_j + \gamma_{1001}x_{ij}Z_j + U_{0j} + U_{1j}X_{ij} + \varepsilon_{ij}. \tag{2.20}$$

Each of these model terms has been defined previously in the chapter: γ_{00}
is the intercept or the grand mean for the model, γ_{10} is the fixed effect of
variable x (vocabulary) on the outcome, U_{0j} represents the random variation
for the intercept across groups, and U_{1j} represents the random variation for
the slope across groups. The additional pieces of the equation in 2.13 are γ_{01}
and γ_{11}. γ_{01} represents the fixed effect of level-2 variable z (average vocabulary) on the outcome. γ_{11} represents the slope for, and value of, the average
vocabulary score for the school. The new term in Model 2.20 is the cross-
level interaction, $\gamma_{1001}x_{ij}z_j$. As the name implies, the cross-level interaction
is simply the interaction between the level-1 and level-2 predictors. In this
context, it is the interaction between an individual's vocabulary score and
the mean vocabulary score for their school. The coefficient for this interaction term, γ_{1001}, assesses the extent to which the relationship between an
examinee's vocabulary score is moderated by the mean for the school that
they attend. A large significant value for this coefficient would indicate that
the relationship between a person's vocabulary test score and their overall
reading achievement is dependent on the level of vocabulary achievement
at their school.

Overview of Three-Level MLMs

With MLMs, it is entirely possible to have three or more levels of data structure. It should be noted that in actual practice, four-level and higher models are rare, however. For our reading achievement data, where the second level was school, the third level might be the district in which the school resides. In that case, we would have multiple equations to consider when expressing the relationship between vocabulary and reading achievement scores, starting at the individual level

$$y_{ijk} = \beta_{0jk} + \beta_{1jk}X_{ijk} + \varepsilon_{ijk}. \tag{2.21}$$

Here, the subscript k represents the level-3 cluster to which the individual belongs. Prior to formulating the rest of the model, we must evaluate if the slopes and intercepts are random at both levels 2 and 3, or only at level 1, for example. This decision should always be based on the theory surrounding the research questions, what is expected in the population, and what is revealed in the empirical data. We will proceed with the remainder of this discussion under the assumption that the level-1 intercepts and slopes are random for both levels 2 and 3, in order to provide a complete description of the most complex model possible when three levels of data structure are present. When the level-1 coefficients are not random at both levels, the terms in the following models for which this randomness is not present would simply be removed. We will address this issue more specifically in Chapter 4, when we discuss the fitting of three-level models using R.

The level-2 and level-3 contributions to the MLM described in 2.13 appear below.

Level 2:

$$\beta_{0jk} = \gamma_{00k} + U_{0jk}$$

$$\beta_{1jk} = \gamma_{10k} + U_{1jk}$$

Level 3:

$$\gamma_{00k} = \delta_{000} + V_{00k}$$

$$\gamma_{10k} = \delta_{100} + V_{10k} \tag{2.22}$$

We can then use simple substitution to obtain the expression for the level-1 intercept and slope in terms of both level-2 and level-3 parameters.

$$\beta_{0jk} = \delta_{000} + V_{00k} + U_{0jk}$$
$$\text{and} \tag{2.23}$$
$$\beta_{1jk} = \delta_{100} + V_{10k} + U_{1jk}$$

In turn, these terms can be substituted into Equation (2.15) to provide the full three-level MLM.

$$y_{ijk} = \delta_{000} + V_{00k} + U_{0jk} + \left(\delta_{100} + V_{10k} + U_{1jk}\right)x_{ijk} + \varepsilon_{ijk}. \tag{2.24}$$

There is an implicit assumption in this expression of 2.24 that there are no cross-level interactions, though these are certainly possible to model across all three levels, or for any pair of levels. In 2.24, we are expressing individuals' scores on the reading achievement test as a function of random and fixed components from the school which they attend, the district in which the school resides, as well as their own vocabulary test score and random variation associated only with themselves. Though not present in 2.24, it is also possible to include variables at both levels 2 and 3, similarly to what we described for the two-level model structure.

Overview of Longitudinal Designs and Their Relationship to MLMs

Finally, we will say just a word about how longitudinal designs can be expressed as MLMs. Longitudinal research designs simply involve the collection of data from the same individuals at multiple points in time. For example, we may have reading achievement scores for examinees in the fall and spring of the school year. With such a design, we would be able to investigate issues around growth scores and change in achievement over time. Such models can be placed in the context of an MLM where the examinee is the level-2 (cluster) variable, and the individual test administration is at level 1. We would then simply apply the two-level model described above, including whichever examinee level variables are appropriate for explaining reading achievement. Similarly, if examinees are nested within schools, we would have a three-level model, with school at the third level, and could apply Model 2.24, once again with whichever examinee- or school-level variables were pertinent to the research question. One unique aspect of fitting longitudinal data in the MLM context is that the error terms can potentially take specific forms that are not common in other applications of multilevel analysis. These error terms reflect the way in which measurements made over time are related to one another, and are typically more complex than the basic error structure that we have described thus far. In Chapter 5 we will look at examples of fitting such longitudinal models with R, and focus much of our attention on these error structures, when each is appropriate, and how they are interpreted. In addition, such MLMs need not take a linear form, but can be adapted to fit quadratic, cubic, or other nonlinear trends over time. These issues will be further discussed in Chapter 5.

Summary

The goal of this chapter was to introduce the basic theoretical underpinnings of multilevel modeling, but not to provide an exhaustive technical discussion of these issues, as there are a number of useful sources available in this regard, which you will find among the references at the end of the text. However, what is given here should stand you in good stead as we move forward with multilevel modeling using R software. We recommend that while reading subsequent chapters you make liberal use of the information provided here, in order to gain a more complete understanding of the output that we will be examining from R. In particular, when interpreting output from R, it may be very helpful for you to come back to this chapter for reminders on precisely what each model parameter means. In the next two chapters we will take the theoretical information from Chapter 2 and apply it to real datasets using two different R libraries, nlme and lme4, both of which have been developed to conduct multilevel analyses with continuous outcome variables. In Chapter 5, we will examine how these ideas can be applied to longitudinal data, and in Chapters 7 and 8, we will discuss multilevel modeling for categorical dependent variables. In Chapter 9, we will diverge from the likelihood-based approaches described here, and discuss multilevel modeling within the Bayesian framework, focusing on application, and learning when this method might be appropriate and when it might not.

3

Fitting Two-Level Models in R

In the previous chapter the multilevel modeling approach to analysis of nested data was introduced, along with relevant notation and definitions of random intercepts and coefficients. We will now devote Chapter 3 to introducing the lme4 package for fitting multilevel models in R. In Chapter 1 we provided an overview of the R lm() function for fitting linear regression models. As will become apparent, the syntax used in the estimation of multilevel models in R is very similar to that of single-level linear models. After providing a brief discussion of the lme4 package for fitting multilevel models for continuous data, we will devote the remainder of the chapter to extended examples applying the principles that we introduced in Chapter 2, using R.

Currently, the main R library for the modeling of multilevel models is lme4, which can be used for fitting basic as well as more advanced multilevel models. Within lme4, the function call to run multilevel models for continuous outcome variables is lmer(). In the following sections of this chapter, we will demonstrate and provide examples of using this package to run basic multilevel models in R. Following is the basic syntax for this function. Details regarding its use and the various options will be provided in the examples.

```
lmer(formula, data = NULL, REML = TRUE, control = lmerControl(),
    start = NULL, verbose = 0L, subset, weights, na.action,
    offset, contrasts = NULL, devFunOnly = FALSE, ...)
```

For simple linear multilevel models, the only necessary R subcommands are the formula (consisting of the fixed and random effects), and the data. The rest of the subcommands can be used to customize models and to provide additional output. This chapter will first focus on the definition of simple multilevel models, and then demonstrate a few options for model customization and assumption checking.

Simple (Intercept-Only) Multilevel Models

In order to demonstrate the use of R for fitting multilevel models, let's return to the example that we introduced in Chapter 2. Specifically, a researcher was interested in determining the extent to which vocabulary scores can be used to predict general reading achievement. Students were nested within schools, so that standard linear regression models would not be appropriate. In this case, school is a random effect, while vocabulary scores are fixed. We will first fit the null model in which there is not an independent variable, but only the intercept. This model is useful for obtaining estimates of the residual and intercept variance when only the clustering by school is considered, as in Equation (2.11). The lmer syntax necessary for estimating the null model appears below.

```
Model3.0 <- lmer(geread~1 +(1|school), data=Achieve)
```

We can obtain output from this model by typing summary(Model3.0).

```
summary(Model3.0)
Linear mixed model fit by REML ['lmerMod']
Formula: geread ~ 1 + (1 | school)
   Data: Achieve

REML criterion at convergence: 46268.3

Scaled residuals:
    Min      1Q  Median      3Q     Max
-2.3229 -0.6378 -0.2138  0.2850  3.8812

Random effects:
 Groups   Name        Variance Std.Dev.
 school   (Intercept) 0.3915   0.6257
 Residual             5.0450   2.2461
Number of obs: 10320, groups:  school, 160

Fixed effects:
            Estimate Std. Error t value
(Intercept)  4.30675    0.05498   78.34
```

Although this is a null model in which there is not an independent variable, it does provide some useful information that will help us understand the structure of the data. In particular, the null model provides estimates of the variance among the individuals σ^2, and among the clusters τ^2. In turn, these values can be used to estimate ρ_I (the ICC), as in Equation (2.5). Here, the value would be

$$\hat{\rho}_I = \frac{0.392}{0.392 + 5.045} = 0.072.$$

We interpret this value to mean that the correlation of reading test scores among students within the same schools is approximately 0.072.

In order to fit a model with vocabulary test score as the independent variable using lmer, we submit the following syntax in R.

```
Model3.1 <- lmer(geread~gevocab + (1|school), data=Achieve)
```

In the first part of the function call we define the formula for the model fixed effects, which is very similar to model definition of linear regression using lm(). The statement geread~gevocab essentially says that the reading score is predicted with the vocabulary score fixed effect. The call in parentheses defines the random effects and the nesting structure. If only a random intercept is desired, the syntax for the intercept is "1." In this example, (1|school) indicates that only a random intercepts model will be used and that the random intercept varies within school. This corresponds to the data structure of students nested within schools. Fitting this model, which is saved in the output object Model3.1, we obtain the following output.

```
summary(Model3.1)
Linear mixed model fit by REML ['lmerMod']
Formula: geread ~ gevocab + (1 | school)
   Data: Achieve

REML criterion at convergence: 43137.2

Scaled residuals:
    Min      1Q  Median      3Q     Max
-3.0823 -0.5735 -0.2103  0.3207  4.4334

Random effects:
 Groups   Name        Variance Std.Dev.
 school   (Intercept) 0.09978  0.3159
 Residual             3.76647  1.9407
Number of obs: 10320, groups:  school, 160

Fixed effects:
             Estimate Std. Error t value
(Intercept) 2.023356   0.049309   41.03
gevocab     0.512898   0.008373   61.26

Correlation of Fixed Effects:
        (Intr)
gevocab -0.758
```

From this summary we obtain parameter estimates, standard errors, and t statistics for testing the statistical significance of the estimates. In addition, we also have the correlation estimate between the fixed-effect slope and the fixed-effect intercept, as well as a brief summary of the model residuals, including the minimum, maximum, and first, second (median), and third quartiles. The correlation of the fixed effects represents the estimated correlation, if we had repeated samples, of the two fixed effects (i.e. the intercept and slope for gevocab). Oftentimes this correlation is not particularly interesting. From this output we can see that gevocab is a statistically significant predictor of geread ($t = 61.26$), and that as vocabulary score increases by 1 point, reading ability increases by 0.513 points. Note that we will revisit the issue of statistical significance when we examine confidence intervals for the model parameter, below. Indeed, we would recommend that researchers rely on the confidence intervals when making such determinations.

In addition to getting estimates of the fixed effects in Model3.1, we can also ascertain how much variation in geread is present across schools. Specifically, the output shows that after accounting for the impact of gevocab, the estimate of variation in intercepts across schools is 0.09978, while the within-school variation is estimated as 3.77647. We can tie these numbers directly back to our discussion in Chapter 2, where $\tau_0^2 = 0.09978$ and $\sigma^2 = 3.76647$. In addition, the overall fixed intercept, denoted as γ_{00} in Chapter 2, is 2.023, and can be interpreted as the mean of geread when the gevocab score is 0.

Finally, it is possible to estimate the proportion of variance in the outcome variable that is accounted for at each level of the model. In Chapter 1, we saw that with single-level OLS regression models, the proportion of response variable variance accounted for by the model is expressed as R^2. In the context of multilevel modeling, R^2 values can be estimated for each level of the model (Snijders and Bosker, 1999). For level 1, we can calculate

$$R_1^2 = 1 - \frac{\sigma_{M1}^2 + \tau_{M1}^2}{\sigma_{M0}^2 + \tau_{M0}^2} = 1 - \frac{3.76647 + .09978}{5.045 + .3915} = 1 - \frac{3.86625}{5.4365} = 1 - .7112 = .2889.$$

This result tells us that level 1 of Model3.1 explains approximately 29% of the variance in the reading score above and beyond that accounted for in the null model. We can also calculate a level-2 R^2:

$$R_2^2 = 1 - \frac{\sigma_{M1}^2 / B + \tau_{M1}^2}{\sigma_{M0}^2 / B + \tau_{M0}^2}$$

where B is the average size of the level-2 units, or schools in this case. R provides us with the number of individuals in the sample, 10,320, and the number of schools, 160, so that we can calculate B as $10{,}320/160 = 64.5$. We can now estimate

$$R_2^2 = 1 - \frac{\sigma_{M1}^2 / B + \tau_{M1}^2}{\sigma_{M0}^2 / B + \tau_{M0}^2} = 1 - \frac{\dfrac{3.76647}{64.5} + .09978}{\dfrac{5.045}{64.5} + .3915} = 1 - \frac{.0583}{.0778} = 1 - .7493 = .2507$$

The model in the previous example was quite simple, only incorporating one level-1 predictor. In many applications, researchers will have predictor variables at both level 1 (student) and level 2 (school). Incorporation of predictors at higher levels of analysis is very straightforward in R, and is done in exactly the same manner as incorporation of level-1 predictors. For example, let's assume that in addition to a student's vocabulary test performance, the researcher also wanted to determine whether the size of the school (senroll) also has a statistically significant impact on overall reading score. In that instance, adding the level-2 predictor, school enrollment, would result in the following R syntax:

```
Model3.2 <- lmer(geread~gevocab+senroll+(1|school),data=Achieve)
summary(Model3.2)
Linear mixed model fit by REML ['lmerMod']
Formula: geread ~ gevocab + senroll + (1 | school)
   Data: Achieve

REML criterion at convergence: 43152.1

Scaled residuals:
    Min      1Q  Median      3Q     Max
-3.0834 -0.5729 -0.2103  0.3212  4.4336

Random effects:
 Groups   Name        Variance Std.Dev.
 school   (Intercept) 0.1003   0.3168
 Residual             3.7665   1.9408
Number of obs: 10320, groups:  school, 160

Fixed effects:
              Estimate Std. Error t value
(Intercept)  2.0748819  0.1140074   18.20
gevocab      0.5128708  0.0083734   61.25
senroll     -0.0001026  0.0002051   -0.50

Correlation of Fixed Effects:
         (Intr) gevocb
gevocab -0.327
senroll -0.901 -0.002
```

Note that in this particular function call, senroll is included only in the fixed part of the model and not in the random part. This variable thus only has a fixed (average) effect and is the same across all schools. We will see shortly how to incorporate a random coefficient in this model.

From these results we can see that, in fact, enrollment did not have a statistically significant relationship with reading achievement ($t = -0.50$). In addition, notice that there were some minor changes in the estimates of the other model parameters, but a fairly large change in the correlation between the fixed effect of the gevocab slope and the fixed effect of the intercept, from −0.758 to −0.327. The slope for senroll and the intercept were very strongly negatively correlated, and the slopes of the fixed effects exhibited virtually no correlation (−0.002). As noted before, these correlations are typically not very informative in terms of understanding the dependent variable, and will therefore rarely be discussed in any detail in reporting analysis results. The R^2 values for levels 1 and 2 appear below.

$$R_1^2 = 1 - \frac{\sigma_{M1}^2 + \tau_{M1}^2}{\sigma_{M0}^2 + \tau_{M0}^2} = 1 - \frac{3.7665 + .1003}{5.045 + .3915} = 1 - \frac{3.8668}{5.4365} = 1 - .7112 = .2887.$$

$$R_2^2 = 1 - \frac{\sigma_{M1}^2 / B + \tau_{M1}^2}{\sigma_{M0}^2 / B + \tau_{M0}^2} = 1 - \frac{\dfrac{3.7665}{64.5 + .1003}}{\dfrac{5.045}{64.5 + .3915}} = 1 - \frac{.0583}{.0778} = 1 - .7494 = .2506$$

These values are nearly identical to those obtained for the model without `senroll`. This is because there was very little change in the variances when `senroll` was included in the model, thus the amount of variance explained at each level was largely unchanged.

Interactions and Cross-Level Interactions Using R

Interactions among the predictor variables, and in particular cross-level interactions, can be very important features in the application of multilevel models. Cross-level interactions occur when the impact of a level-1 variable on the outcome (e.g. vocabulary score) differs depending on the value of the level-2 predictor (e.g. school enrollment). Interactions, be they within the same level or cross-level, are simply the product of two predictors. Thus, incorporation of interactions and cross-level interactions in multilevel modeling is accomplished in much the same manner as we saw for the `lm()` function in Chapter 1. Following are examples for fitting an interaction model for two level-1 variables (Model3.3) and a cross-level interaction involving level-1 and level-2 variables (Model3.4).

```
Model3.3 <- lmer(geread~gevocab + age + gevocab*age +
(1|school), data=Achieve)
```

```
Model3.4 <- lmer(geread~gevocab + senroll + gevocab*senroll +
(1|school), data=Achieve)
```

Model3.3 defines a multilevel model where two level-1 (student level) predictors interact with one another. Model3.4 defines a multilevel model with a cross-level interaction: where a level-1 (student level) and a level-2 (school level) predictor are interacting. As can be seen, there is no difference in the treatment of variables at different levels when computing interactions.

```
Model3.3 <- lmer(geread~gevocab+age+gevocab*age +(1|school),
data=Achieve)
```

```
summary(Model3.3)
Linear mixed model fit by REML ['lmerMod']
Formula: geread ~ gevocab + age + gevocab * age + (1 | school)
   Data: Achieve

REML criterion at convergence: 43143.5

Scaled residuals:
    Min      1Q  Median      3Q     Max
-3.0635 -0.5706 -0.2108  0.3191  4.4467

Random effects:
 Groups    Name        Variance Std.Dev.
 school    (Intercept) 0.09875  0.3143
 Residual              3.76247  1.9397
Number of obs: 10320, groups:  school, 160

Fixed effects:
              Estimate Std. Error t value
(Intercept)   5.187208   0.866786   5.984
gevocab      -0.028077   0.188145  -0.149
age          -0.029368   0.008035  -3.655
gevocab:age   0.005027   0.001750   2.873

Correlation of Fixed Effects:
            (Intr) gevocb age
gevocab     -0.879
age         -0.998  0.879
gevocab:age  0.877 -0.999 -0.879
```

Looking at the output from Model3.5, both age ($t = -3.65$) and the interaction (gevocab:age) between age and vocabulary ($t = 2.87$) are statistically significant predictors of reading. Focusing on the interaction, the sign on the coefficient is positive indicating an enhancing effect: as age increases, the relationship between reading and vocabulary becomes stronger. Interestingly, when both age and the interaction are included in the model, the relationship between vocabulary score and reading performance is no longer statistically significant.

Next, let's examine a model that includes a cross-level interaction.

```
Model3.4 <- lmer(geread~gevocab+senroll+gevocab*senroll +
(1|school), data=Achieve)
```

```
summary(Model3.4)
Linear mixed model fit by REML ['lmerMod']
Formula: geread ~ gevocab + senroll + gevocab * senroll + (1 |
school)
   Data: Achieve

REML criterion at convergence: 43163.6

Scaled residuals:
    Min       1Q  Median       3Q      Max
-3.1228  -0.5697  -0.2090   0.3188   4.4359

Random effects:
 Groups    Name           Variance Std.Dev.
 school    (Intercept)    0.1002   0.3165
 Residual                 3.7646   1.9403
Number of obs: 10320, groups:  school, 160

Fixed effects:
                    Estimate Std. Error t value
(Intercept)        1.748e+00  1.727e-01  10.118
gevocab            5.851e-01  2.986e-02  19.592
senroll            5.121e-04  3.186e-04   1.607
gevocab:senroll   -1.356e-04  5.379e-05  -2.520

Correlation of Fixed Effects:
             (Intr) gevocb senrll
gevocab      -0.782
senroll      -0.958  0.735
gevcb:snrll   0.752 -0.960 -0.766
```

The output from Model3.4 has a similar interpretation to that of Model3.3. In this example, when school enrollment is used instead of age as a predictor, the main effect of vocabulary ($t = 19.59$) and the interaction between vocabulary and school enrollment ($t = -2.51$) are statistically significant predictors of reading achievement. Given that the sign on the interaction coefficient is negative, we would conclude that there is a buffering or inhibitory effect. In other words, as the size of the school increases, the relationship between vocabulary and reading achievement becomes weaker.

Random Coefficients Models using R

In Chapter 2 we described the random coefficients model, in which the impact of the independent variable on the dependent is allowed to vary across the level-2 effects. When defining random effects, as mentioned above, 1 stands for the intercept, so that if all we desire is a random intercepts model, as in the previous example, the syntax (1|school) is sufficient. If, however, we want to allow a level-1 slope to randomly vary across level-2 units, we will need to change this part of the syntax. For example, let's alter Model3.1, where we predicted geread from gevocab by allowing the effect of gevocab to be random. Incorporating such random coefficient effects into a multilevel model using lmer occurs in the random part of the model syntax.

```
Model 3.5 <- lmer(geread~gevocab + (gevocab|school),
data=Achieve).¹
```

Notice that we are no longer explicitly stating in the specification that there is a random intercept. Once a random slope is defined, the random intercept becomes implicit, so we no longer need to specify it (i.e. it is included by default). If we do not want the random intercept while modeling the random coefficient, we would include a -1 immediately prior to gevocab. The random slope and intercept syntax will generate the following model summary:

```
Model3.5 <- lmer(geread~gevocab + (gevocab|school),
data=Achieve)

summary(Model3.5)
Linear mixed model fit by REML ['lmerMod']
Formula: geread ~ gevocab + (gevocab | school)
   Data: Achieve

REML criterion at convergence: 42992.9

Scaled residuals:
    Min      1Q  Median      3Q     Max
-3.7102 -0.5674 -0.2074  0.3176  4.6775

Random effects:
 Groups   Name        Variance Std.Dev. Corr
 school   (Intercept) 0.1025   0.3202
          gevocab     0.0193   0.1389   0.52
 Residual             3.6659   1.9147
Number of obs: 10320, groups:  school, 160

Fixed effects:
            Estimate Std. Error t value
(Intercept)  4.34411    0.03271   132.8
gevocab      0.52036    0.01442    36.1

Correlation of Fixed Effects:
        (Intr)
gevocab 0.362
```

An examination of the results shows that gevocab is statistically signif-icantly related to geread, across schools. The estimated coefficient, 0.520, corresponds to γ_{10}, and is interpreted as the average impact of the predictor on the outcome across schools. In addition, the value .0193 is the estimate of τ_1^2 and reflects the variation in coefficients across schools. A relatively large value of this estimate indicates that the coefficient varies from one school to another; i.e. the relationship of the independent and dependent variables dif-fers across schools. As before, we also have the estimates of τ_0^2 (.1025) and σ^2 (3.6659). Taken together, these results show that the largest source of random variation in geread is among students within schools, with lesser variation coming from differences in the conditional mean (intercept) and the coeffi-cient for gevocab across schools. At the end of this chapter, we will see how to obtain confidence intervals for the fixed and random model effects. These can be used to make determinations regarding the statistical significance of each of these terms.

A model with two random slopes can be defined in much the same way as for a single slope. As an example, suppose that our researcher is inter-ested in determining whether the age of the student also impacted read-ing performance, and wants to allow this effect to vary from one school to another. In lme4, random effects are allowed to be either correlated or uncorrelated, providing excellent modeling flexibility. As an example, refer to Models 3.6 and 3.7, each of which predicts geread from gevocab and age allowing both gevocab and age to be random across schools. They differ, however, in that with Model3.6 the random slopes are treated as cor-related with one another, whereas in Model3.7 they are specified as being uncorrelated. This lack of correlation in Model3.7 is expressed by having the separate random-effect terms (gevocab|school) and (age|school), while in contrast, Model3.6 includes both random effects in the same term (gevocab + age|school).

```
Model3.6 <- lmer(geread~gevocab + age+(gevocab + age|school),
Achieve)¹


Model3.7 <- lmer(geread~gevocab + age+ (gevocab|school) +
(age|school), Achieve)¹


Model3.6 <- lmer(geread~gevocab + age +(gevocab + age|school),
data=Achieve)
summary(Model3.6)
Linear mixed model fit by REML ['lmerMod']
Formula: geread ~ gevocab + age + (gevocab + age | school)
   Data: Achieve

REML criterion at convergence: 42995.3
```

```
Scaled residuals:
    Min      1Q  Median      3Q     Max
-3.6735 -0.5682 -0.2091  0.3184  4.6840
```

```
Random effects:
 Groups   Name        Variance  Std.Dev. Corr
 school   (Intercept) 1.022e-01 0.319700
          gevocab     1.902e-02 0.137918  0.53
          age         2.509e-05 0.005009 -0.28 -0.96
 Residual             3.664e+00 1.914181
Number of obs: 10320, groups:  school, 160
```

```
Fixed effects:
             Estimate Std. Error t value
(Intercept)  4.343872   0.032686 132.895
 gevocab     0.519277   0.014350  36.187
 age        -0.008882   0.003822  -2.324
```

```
Correlation of Fixed Effects:
        (Intr) Cgevcb
 gevocab  0.367
 age     -0.020 -0.048
```

```
Model3.7 <- lmer(geread~gevocab + age +(gevocab|school) +
(age|school), data=Achieve)
```

```
summary(Model3.7)
Linear mixed model fit by REML ['lmerMod']
Formula: geread ~ gevocab + age + (gevocab | school) + (age |
school)
   Data: Achieve
```

```
REML criterion at convergence: 42996.5
```

```
Scaled residuals:
    Min      1Q  Median      3Q     Max
-3.6937 -0.5681 -0.2081  0.3182  4.6744
```

```
Random effects:
 Groups     Name        Variance  Std.Dev.  Corr
 school     (Intercept) 2.914e-02 0.1707120
            gevocab     1.919e-02 0.1385422 1.00
 school.1   (Intercept) 7.272e-02 0.2696677
            age         7.522e-07 0.0008673 1.00
 Residual               3.665e+00 1.9143975
Number of obs: 10320, groups:  school, 160
```

```
Fixed effects:
             Estimate Std. Error t value
(Intercept)  4.343789   0.032644 133.065
```

```
gevocab      0.519190    0.014397   36.063
age         -0.008956    0.003798   -2.358
```

```
Correlation of Fixed Effects:
         (Intr)  gevcb
gevocab 0.368
age      0.015  0.033
```

Notice the difference in how random effects are expressed in lmer between Models 3.6 and 3.7. With random effects, R reports estimate for the variability of the random intercept, variability for each random slope, and the correlations between the random intercept and the random slopes. Output in Model3.7, however, reports two different sets of uncorrelated random effects with an intercept estimated for each separate random effect. If uncorrelated random effects are desired without the separate intercept estimated for each, the following code can be used to eliminate the extra intercepts.

```
Model3.6a <- lmer(geread~gevocab + Cage + (1|school) + (-1 +
gevocab|school) + (-1 + age|school), data=Achieve)
```

```
summary(Model3.6a)
Linear mixed model fit by REML ['lmerMod']
Formula: geread ~ gevocab + age + (1 | school) + (-1 + gevocab
| school) +
    (-1 + age | school)
   Data: Achieve
```

```
REML criterion at convergence: 43014.1
```

```
Scaled residuals:
    Min      1Q  Median      3Q     Max
-3.6084 -0.5664 -0.2058  0.3087  4.5437
```

```
Random effects:
 Groups    Name        Variance Std.Dev.
 school    (Intercept) 0.09547  0.3090
 school.1  gevocab     0.01833  0.1354
 school.2  age         0.00000  0.0000
 Residual              3.66719  1.9150
Number of obs: 10320, groups:  school, 160
```

```
Fixed effects:
             Estimate Std. Error t value
(Intercept)  4.351259   0.032014 135.917
 gevocab     0.521053   0.014133  36.869
 age        -0.008652   0.003807  -2.272
```

```
Correlation of Fixed Effects:
         (Intr)  gevcb
gevocab 0.036
age      0.003  0.034
```

Centering Predictors

As per the discussion in Chapter 2, it may be advantageous to center predictors, especially when interactions are incorporated. Centering predictors can provide easier interpretation of interaction terms as well as help alleviating issues of multicollinearity arising from inclusion of both main effects and interactions in the same model. Recall that centering of a variable entails the subtraction of a mean value from each score on the variable. Centering of predictors can be accomplished through R by the creation of new variables. For example, returning to Model3.3, grand mean centered gevocab and age variables can be created with the following syntax:

```
Cgevocab <- Achieve$gevocab - mean(Achieve$gevocab)
Cage <- Achieve$age - mean(Achieve$age)
```

Once mean centered versions of the predictors have been created, they can be incorporated into the model in the same manner as before.

```
Model3.3C <-lmer(geread~Cgevocab + Cage + Cgevocab*Cage +
(1|school), data=Achieve)

summary(Model3.3C)
Linear mixed model fit by REML ['lmerMod']
Formula: geread ~ Cgevocab + Cage + Cgevocab * Cage + (1 |
school)
   Data: Achieve

REML criterion at convergence: 43143.5

Scaled residuals:
    Min      1Q  Median      3Q     Max
-3.0635 -0.5706 -0.2108  0.3191  4.4467

Random effects:
 Groups   Name        Variance Std.Dev.
 school   (Intercept) 0.09875  0.3143
 Residual             3.76247  1.9397
Number of obs: 10320, groups:  school, 160
```

```
Fixed effects:
                Estimate Std. Error t value
(Intercept)     4.332327   0.032062 135.124
Cgevocab        0.512480   0.008380  61.159
Cage           -0.006777   0.003917  -1.730
Cgevocab:Cage   0.005027   0.001750   2.873

Correlation of Fixed Effects:
            (Intr) Cgevcb Cage
Cgevocab    0.008
Cage        0.007  0.053
Cgevocab:Cg 0.043  0.021  0.205
```

Focusing on the fixed effects of the model, there are some changes in their values. These differences are likely due to the presence of multicollinearity issues in the original uncentered model. The interaction is still significant ($t = 2.87$); however, there is now a significant effect of vocabulary ($t = 61.15$), and age is no longer a significant predictor ($t = -1.73$). Focusing on the interaction, recall that when predictors are centered, the interaction can be interpreted as the effect of one variable while holding the second variable constant. Thus, since the sign on the interaction is positive, if we hold age constant, vocabulary has a positive impact on reading ability.

Additional Options

Parameter Estimation Method

By default, lme4 uses restricted maximum likelihood estimation (REML). However, it also allows for the use of maximum likelihood estimation (ML) instead. Model3.8 demonstrates syntax for fitting a multilevel model using ML. In order to change the estimation method, the call is REML = FALSE.

```
Model3.8 <- lmer(geread~gevocab + (1|school), data=Achieve,
REML=FALSE)
```

Estimation Controls

Sometimes a correctly specified model will not reach a solution (converge) using the default settings for model convergence. Many times, this problem can be fixed by changing the default estimation controls using the control option. Quite often, convergence issues can be fixed by changing the model iteration limit (maxIter), or by changing the model optimizer (opt). In order to specify which controls will be changed, R must be given a list of controls and their new values. For example, control=list(maxIter=100, opt="optim") would change the maximum number of iterations to 100 and the optimizer to "optim." These control options are placed in the R code in the same manner as choice of estimation method (separated from the rest of the syntax with a comma). A comprehensive list of estimation controls can be found on the R help `?lme4` pages.

```
Model3.8 <- lmer(geread~gevocab + (1|school), data=Achieve,
REML=FALSE, control=list(maxIter=100, opt="optim"))
```

Comparing Model Fit

Often, one wishes to compare the fit of multiple models in order to identify a best fitting model given our data. Model comparison information can be obtained through use of the anova() function. This function can be used to provide two different types of model fit information: AIC and BIC statistics, and the chi-square test of model fit. When we are working with nested models, where one model is a more constrained (i.e. simpler) version of another, we may wish to test whether overall fit of the two models differs significantly (as opposed to using AIC and BIC statistics, which are more general model comparison statistics and can't provide this level of detail). Such hypothesis testing is possible using the chi-square difference test based on the deviance statistic. When the fits of nested models are being compared, the difference in chi-square values for each model deviance can be used to compare model fit. It is important to note that for the chi-square test of deviance to be accurate for multilevel models, maximum likelihood estimation must be used. When maximum likelihood is used, both fixed and random effects are compared simultaneously. When restricted maximum likelihood is used, only random effects are compared. The anova() command in lme4 automatically refits models using maximum likelihood if it was not run with this estimation previously.

Let us return to the first two models in this chapter. If you recall, we began by running a null model predicting geread from no predictors and a simple random intercepts model predicting geread from gevocab. We can compare the fit for these two models by using the function call anova(Model3.0,Model3.1). We obtain the following output.

```
anova(Model3.0,Model3.1)
refitting model(s) with ML (instead of REML)
Data: Achieve
Models:
Model3.0: geread ~ 1 + (1 | school)
Model3.1: geread ~ gevocab + (1 | school)
         Df    AIC    BIC logLik deviance  Chisq Chi Df
Pr(>Chisq)
Model3.0  3 46270 46292 -23132    46264
Model3.1  4 43132 43161 -21562    43124 3139.9      1  <
2.2e-16 ***
---
Signif. codes:  0 '***' 0.001 '**' 0.01 '*' 0.05 '.' 0.1 ' ' 1
```

Referring to the AIC and BIC statistics, recall that smaller values reflect better model fit. For Model3.1, the AIC and BIC are 43132 and 43161, respectively, whereas for Model3.0 the AIC and BIC were 46270 and 46292. Given that the values for both statistics are smaller for Model3.1, we would conclude that it provides a better fit to the data. Since these models are nested, we may also look at the chi-square test for deviance. This test yielded a statistically significant chi-square ($\chi^2 = 3139.9$, $p < .001$) indicating that Model3.1 provided a significantly better fit to the data than does Model3.0. Substantively, this means that we should include the predictor variable `geread`, which results of the hypothesis test also supported.

lme4 and Hypothesis Testing

It is hard to overlook one nuance of the lme4 library: the lack of reported *p*-values for the estimated effects. The issue of *p*-values in multilevel modeling is somewhat complex, with a full discussion lying outside the bounds of this book. However, we should note that the standard approach for finding *p*-values based on using the reference *t* distribution, which would seem to be the intuitively correct thing to do, does in fact not yield correct values in many instances. Therefore, some alternative approach for obtaining them is necessary. Douglas Bates, the developer of `lme4`, recommends the use of bootstrapped confidence intervals in order to accurately estimate the significance of fixed and random effects in multilevel models.

In R, the `confint()` function and the following code can be used to estimate bootstrapped confidence intervals around the fixed and random effects for Model3.5. There are a number of options using the `confint` function. In particular, when we consider bootstrap intervals, three options include percentile bootstrap (`perc`), the standard error bootstrap (`basic`), and the normal bootstrap (`norm`). Each of these methods relies on the random resampling of *B* samples with replacement, with the default being 500. For the percentile intervals, the multilevel model is fit to each of the *B* samples, and the parameter estimates are saved. The 95% confidence interval for a given estimate corresponds to the 2.5th and 97.5th percentiles of the bootstrap distribution. For the standard error bootstrap, the standard deviation of the *B* samples is calculated and serves as the standard error estimate. The confidence interval for a given parameter is then estimated as:

$$\theta \pm z_\alpha SE_B \qquad\qquad (3.1)$$

where

θ = parameter estimate as calculated from original data
z_α = critical value from the standard normal distribution
SE_B = bootstrap standard error; i.e. standard deviation of the bootstrap sample

The normal bootstrap is very similar to the standard error approach, the difference being that the B samples are generated from a normal distribution with marginal statistics (i.e. means, variances) equal to those of the raw data, rather than through resampling. In other respects, it is equivalent to the standard error method. Two other approaches for calculating confidence intervals are available using confint. The first of these applies only to the fixed effects, and is based on the assumption of normally distributed errors. The confidence interval is then calculated as

$$b_{x1} \pm z_{\alpha} SE_{MLE} \tag{3.2}$$

where
b_{x1} = coefficient linking X1 to Y
z_{α} = critical value from the standard normal distribution
SE_B = MLE standard error

The final confidence interval method that we can use is called the profile confidence interval, and is not based on any assumptions about the distribution of the errors. Instead, we select as the bounds of the confidence interval two points on either side of the MLE estimate with likelihood equal to

$$\text{MLE} - 0.5 * (1 - \alpha) \text{ percentile of the Chi} - \text{square distribution with DF} = 1.$$

In order to construct the confidence interval, we find the estimates of the parameter of interest (e.g. random intercept variance) that yield a non-significant likelihood ratio test of the null hypothesis $H_0 : \theta = \theta_0$, where θ_0 is an estimate of the parameter. Thus, the profile interval consists of all the values of θ_0 that yield a non-significant test of this null hypothesis. Put another way, compute the likelihood $(\ln L_1(\theta_0))$ for each value of θ_0 within some wide range of values (e.g. 100 equidistant points). The lower bound of the interval is the smallest θ_0 value for which $\ln L_1(\theta_0) \geq \ln L_1(\theta^*) - 0.5 * \chi^2_{cv}$, with 1 degree of freedom. If we set $\alpha = 0.05$, then $\chi^2_{cv} = 3.84$, so that $0.5 * \chi^2_{cv} = 1.92$.

These confidence intervals can be obtained using the confint function, as below.

```
Model3.5 <- lmer(geread~gevocab + (gevocab|school),
data=Achieve)
summary(Model3.5)
confint(Model3.5, method=c("boot"), boot.type=c("perc"))
confint(Model3.5, method=c("boot"), boot.type=c("basic"))
confint(Model3.5, method=c("boot"), boot.type=c("norm"))
confint(Model3.5, method=c("Wald"))
confint(Model3.5, method=c("profile"))

confint(Model3.5, method=c("boot"), boot.type=c("perc"))
Computing bootstrap confidence intervals ...
```

```
                2.5 %      97.5 %
.sig01      0.2664702 0.3812658
.sig02      0.2798048 0.7387527
.sig03      0.1105445 0.1644085
.sigma      1.8879782 1.9408446
(Intercept) 4.2844713 4.4028048
Cgevocab    0.4913241 0.5511338
confint(Model3.5, method=c("boot"), boot.type=c("basic"))
Computing bootstrap confidence intervals ...
                2.5 %      97.5 %
.sig01      0.2638048 0.3767120  .
.sig02      0.3131177 0.7405494
.sig03      0.1148660 0.1662661
.sigma      1.8900770 1.9412491
(Intercept) 4.2795679 4.4078229
Cgevocab    0.4909470 0.5484041
confint(Model3.5, method=c("boot"), boot.type=c("norm"))
Computing bootstrap confidence intervals ...
                2.5 %      97.5 %
.sig01      0.2633918 0.3799952
.sig02      0.3302355 0.7388453
.sig03      0.1140536 0.1655553
.sigma      1.8871793 1.9429654
(Intercept) 4.2830980 4.4057360
Cgevocab    0.4935254 0.5488512
confint(Model3.5, method=c("Wald"))
                2.5 %      97.5 %
.sig01          NA         NA
.sig02          NA         NA
.sig03          NA         NA
.sigma          NA         NA
(Intercept) 4.2799887 4.4082236
Cgevocab    0.4921036 0.5486107
confint(Model3.5, method=c("profile"))
Computing profile confidence intervals ...
                2.5 %      97.5 %
.sig01      0.2623302 0.3817036
.sig02      0.2926177 0.7139976
.sig03      0.1141790 0.1656869
.sigma      1.8884541 1.9414934
(Intercept) 4.2790205 4.4083316
Cgevocab    0.4920097 0.5490748
```

The last section of the output contains the confidence intervals for the fixed and random effects of Model3.5. The top three rows correspond to the random intercept (.sig01), the correlation between the random intercept and random slope (.sig02), and the random slope (.sig03). According to these confidence intervals, since 0 does not appear in the interval, there is statistically significant variation in the intercept across schools CI95[.262, .382], the slope for gevocab across schools CI95[.114, .166], and there is a significant relationship between the random slope and the random intercept CI95[.293, .713].

The bottom two rows correspond to the fixed effects. According to the confidence intervals, the intercept (γ_{00}) is significant CI95[4.279, 4.408] and the slope for gevocab is significant CI95[.492, .549], each corresponding with the *t* values presented in the output previously displayed in the chapter.

Summary

In Chapter 3, we put the concepts that we learned in Chapter 2 to work using R. We learned the basics of fitting two-level models when the dependent variable is continuous, using the lme4 package. Within this multilevel framework, we learned how to fit the null model, as well as the random intercept and random slopes models. We also included independent variables at both levels of the data, and finally, learned how to compare the fit of models with one another. This last point will prove particularly useful as we engage in the process of selecting the most parsimonious model (i.e. the simplest) that also adequately explains the dependent variable. Of greatest import in this chapter, however, is understanding how to fit multilevel models using lme4 in R, and correctly interpreting the resultant output. If you have mastered those skills, then you are ready to move to Chapter 4, where we extend the model to include a third level in the hierarchy. As we will see, the actual fitting of these three-level models is very similar to that for the two-level models we have studied here.

Note

1. In these models, gevocab and age were grand mean centered to fix convergence issues.

4

Three-Level and Higher Models

In Chapters 2 and 3, we introduced the multilevel modeling framework, and demonstrated the use of the lme4 R package for fitting two-level models. In Chapter 4, we will expand upon this basic two-level framework by fitting models with additional levels of data structure. As described in Chapter 2, it is conceivable for a level-1 unit, such as students, to be nested in higher-level units, such as classrooms. Thus, in keeping with our examples, we might assume that at least a portion of a student's performance on a reading test is due to the classroom in which they find themselves. Each classroom may have a unique learning context, which in turn would contribute to the performance of the students in the classroom. These unique contexts include such things as the quality of the teacher, the presence of disruptive students, and the time of day in which students are in the class, among others. Furthermore, as we saw in Chapters 2 and 3, the impact of fixed effects on the dependent variable can vary among the level-2 units, resulting in a random slopes model. Now we will see that, using R, it is possible to estimate models with three or more levels of a nested structure, and that the R command for defining and fitting these models is very similar to that which we used in the two-level case. Within the lme4 package, the same function call that we used for two-level models can be used to define models with three or more levels:

```
lmer(formula, data = NULL, REML = TRUE, control = lmer-
Control(),   start = NULL, verbose = 0L, subset, weights,
na.action,   offset, contrasts = NULL, devFunOnly = FALSE, ...)
```

In this chapter we will continue working with the Achieve data that were described in Chapter 3. In our Chapter 3 examples, we included two levels of data structure, students within schools, along with associated predictors of reading achievement at each level. We will now add a third level of structure, classroom, which is nested within schools. In this context, then, students are nested within classrooms, which are in turn nested within schools.

Defining Simple Three-Level Models Using the lme4 Package

The R syntax for defining and fitting models incorporating more than two levels of data structure is very similar to that for two-level models that we have already seen in Chapter 3. Let's begin by defining a null model for prediction of student reading achievement, where regressors might include student-level characteristics, classroom-level characteristics, and school-level characteristics. The syntax to fit a three-level null model appears below, with the results being stored in the object Model4.0.

```
Model4.0 <- lmer(geread~1+(1|school/class), data=Achieve)
```

As can be seen, the syntax for fitting a random intercepts model with three levels is very similar to that for the random intercepts model with two levels. In order to define a model with more than two levels, we need to include the variables denoting the higher levels of the nesting structures, here school (school-level influence) and class (classroom-level influence), and designate the nesting structure of the levels (students within classrooms within schools). Nested structure in lmer is defined as A/B, where A is the higher-level data unit (e.g. school) and B is the lower-level data unit (e.g. classroom). The intercept (1) is denoted as a random effect by its inclusion in the parentheses.

```
Model4.0 <- lmer(geread~1+(1|school/class), data=Achieve)

summary(Model4.0)
Linear mixed model fit by REML ['lmerMod']
Formula: geread ~ 1 + (1 | school/class)
   Data: Achieve

REML criterion at convergence: 46146

Scaled residuals:
    Min      1Q  Median      3Q     Max
-2.3052 -0.6290 -0.2094  0.3049  3.8673

Random effects:
 Groups       Name        Variance Std.Dev.
 class:school (Intercept) 0.2727   0.5222
 school       (Intercept) 0.3118   0.5584
 Residual                 4.8470   2.2016
Number of obs: 10320, groups:  class:school, 568; school, 160

Fixed effects:
            Estimate Std. Error t value
(Intercept)  4.30806    0.05499   78.34
```

As this is a random intercept-only model, there is not much to be interpreted. There are, however, some new pieces of information to take note of. For example, we see two different sets of random effects: random effects for school, so that the intercept is modeled to vary across schools, and random effects for class:school (read, class in school), so that the intercept is modeled to vary across classrooms within schools. Remember from our discussion in Chapter 2 that we can interpret these random intercepts as means of the dependent variable (reading) varying across levels of the random effects (classrooms and schools). We should also note that at the end of the output, R summarizes the sample size for each of the higher-level units. This is a good place to check and be sure that our model is being defined properly, and that the appropriate data are being used. For example with these data, there are multiple classrooms within each school, so it makes sense that we should have a smaller number of schools (school = 160) and a larger number of classrooms (class:school = 568).

Finally, just as we saw in Chapter 3, the confint() function can be used to obtain inference information about our fixed and random effects.

```
confint(Model4.0, method=c("profile"))[1]
                 2.5 %     97.5 %
.sig01       0.4516596 0.5958275
.sig02       0.4658891 0.6561674
.sigma       2.1710519 2.2328459
(Intercept) 4.1997929 4.4160128
```

In this output, .sig01 corresponds to the random intercept across schools, .sig02 corresponds to the random intercept across classrooms within schools, sigma refers to the residual variance, and Intercept provides the confidence interval for the fixed intercept effect (γ_{00}). From these results, we can infer that mean reading scores differ across schools, given that the 95% confidence interval for .sig01 does not include 0. We would reach a similar inference regarding classroom nested within school, because again, the 95% confidence interval for .sig02 does not include 0.

With this knowledge of how to define the higher-level data structure, adding predictors to the fixed portion of a multilevel model with three or more levels is done in exactly the same manner as for a two-level model. For example, we may wish to extend the intercept-only model described above so that it includes several independent variables, such as a student's vocabulary test score (gevocab), the size of the student's reading classroom (clenroll), and the size of the student's school (cenroll). In lmer, the R command for fitting this model and viewing the resultant output would be:

```
Model4.1 <- lmer(geread~gevocab+clenroll+cenroll+(1|school/
class), data=Achieve)

summary(Model4.1)
Linear mixed model fit by REML ['lmerMod']
Formula: geread ~ gevocab + clenroll + cenroll + (1 | school/
class)
   Data: Achieve
```

```
REML criterion at convergence: 43130.9

Scaled residuals:
    Min      1Q   Median      3Q      Max
-3.2212 -0.5673 -0.2079  0.3184  4.4736

Random effects:
 Groups         Name         Variance Std.Dev.
 class:school (Intercept) 0.09047  0.3008
 school       (Intercept) 0.07652  0.2766
 Residual                  3.69790  1.9230
Number of obs: 10320, groups:  class:school, 568; school, 160

Fixed effects:
              Estimate Std. Error t value
(Intercept)  1.675e+00  2.081e-01   8.050
gevocab      5.076e-01  8.427e-03  60.233
clenroll     1.899e-02  9.559e-03   1.986
cenroll     -3.721e-06  3.642e-06  -1.022

Correlation of Fixed Effects:
         (Intr) gevocb clnrll
gevocab  -0.124
clenroll -0.961 -0.062
cenroll  -0.134  0.025 -0.007
```

We can see from the output for Model4.1 that the student's vocabulary score ($t = 60.23$) and size of the classroom ($t = 1.99$) are statistically significantly positive predictors of student reading achievement score, but the size of the school ($t = -1.02$) does not significantly predict reading achievement. As a side note, the significant positive relationship between size of the classroom and reading achievement might seem to be a bit confusing, suggesting that students in larger classrooms had higher reading achievement test scores. However, in this particular case, larger classrooms very frequently included multiple teacher's aides, so that the actual adult to student ratio might be lower than in classrooms with fewer students. In addition, estimates for the random intercepts of classroom nested in school and school have decreased in value from those of the null model, suggesting that when we account for the three fixed effects, some of the mean differences between schools and between classrooms are accounted for.

Confidence intervals for the model effects can be obtained with the con-fint() function.

```
confint(Model4.1, method=c("profile"))[1]
                 2.5 %        97.5 %
.sig01     2.312257e-01  3.660844e-01
.sig02     2.055494e-01  3.398754e-01
.sigma     1.896242e+00  1.950182e+00
```

```
(Intercept)    1.268352e+00  2.082537e+00
gevocab        4.910305e-01  5.244195e-01
clenroll       2.564126e-04  3.768655e-02
cenroll       -1.084902e-05  3.394918e-06
```

In terms of the fixed effects, the 95% confidence intervals demonstrate that vocabulary score and class size are statistically significant predictors of reading score, but school size is not. In addition, we see that although the variation in random intercepts for schools and classrooms nested in schools declined with the inclusion of the fixed effects, we would still conclude that the random intercept terms are different from 0 in the population, indicating that mean reading scores differ across schools, and across classrooms nested within schools.

The R^2 value for Model4.1 can be calculated as:

$$R_1^2 = 1 - \frac{\sigma_{M1}^2 + \tau_{M1}^2}{\sigma_{M0}^2 + \tau_{M0}^2} = 1 - \frac{3.6979 + .09047}{4.847 + .2727} = 1 - \frac{3.78837}{5.1197} = 1 - .73996 = .26.$$

From this value, we see that inclusion of the classroom and school enrollment variables, along with student vocabulary score, results in a model that explains approximately 26% of the variance in the reading score, above and beyond the null model.

We can also ascertain whether including the predictor variables resulted in a better fitting model, as compared to the null model. As we saw in Chapter 3, we can compare model by examining the AIC and BIC values for each, where lower values indicate better fit. Using the anova() command, we can compare Model4.0 (the null model) with Model4.1.

```
anova(Model4.0, Model4.1)
Data: Achieve
Models:
Model4.0: geread ~ 1 + (1 | school/class)
Model4.1: geread ~ gevocab + clenroll + cenroll + (1 | school/
class)
         Df    AIC    BIC  logLik  deviance  Chisq  Chi Df
Pr(>Chisq)
Model4.0  4  46150  46179  -23071    46142
Model4.1  7  43101  43152  -21544    43087  3054.6     3    <
2.2e-16 ***
---
Signif. codes:  0 '***' 0.001 '**' 0.01 '*' 0.05 '.' 0.1 ' ' 1
```

For the original null model, AIC and BIC were 46,150 and 46,179, respectively, which are both larger than the AIC and BIC for Model4.1. Therefore, we would conclude that this latter model including a single predictor variable at each level provides better fit to the data, and thus is preferable to the null model with no predictors. We could also look at the chi-square deviance

test since these are nested models. The chi-square test is statistically significant, indicating Model4.1 is a better fit to the data.

Using lmer, it is very easy to include both single-level and cross-level
interactions in the model once the higher-level structure is understood. For
example, we may have a hypothesis stating that the impact of vocabulary
score on reading achievement varies depending upon the size of the school
that a student attends. In order to test this hypothesis, we will need to include
the interaction between vocabulary score and size of the school, as is done
in Model4.2 below.

```
Model4.2 <- lmer(geread~gevocab+clenroll+cenroll+gevocab*cenro
ll+(1|school/class), data=Achieve)

summary(Model4.2)
Linear mixed model fit by REML ['lmerMod']
Formula: geread ~ gevocab + clenroll + cenroll + gevocab *
cenroll + (1 |
    school/class)
  Data: Achieve

REML criterion at convergence: 43151.8

Scaled residuals:
    Min      1Q  Median      3Q     Max
-3.1902 -0.5683 -0.2061  0.3183  4.4724

Random effects:
 Groups        Name         Variance Std.Dev.
 class:school (Intercept) 0.08856  0.2976
 school       (Intercept) 0.07513  0.2741
 Residual                  3.69816  1.9231
Number of obs: 10320, groups:  class:school, 568; school, 160

Fixed effects:
                  Estimate Std. Error t value
(Intercept)      1.752e+00  2.100e-01    8.341
gevocab          4.900e-01  1.168e-02   41.940
clenroll         1.880e-02  9.512e-03    1.977
cenroll         -1.316e-05  5.628e-06   -2.337
gevocab:cenroll  2.340e-06  1.069e-06    2.190

Correlation of Fixed Effects:
            (Intr) gevocb clnrll cenrll
gevocab     -0.203
clenroll    -0.949 -0.041
cenroll     -0.212  0.542  0.000
gevcb:cnrll  0.166 -0.693 -0.007 -0.766
```

In this example we can see that, other than the inclusion of the higher-level nesting structure in the random effects line, defining a cross-level interaction in a model with more than two levels is no different than was the case for the two-level models that we worked with in Chapter 3. In terms of hypothesis testing results, we find that student vocabulary ($t = 41.94$) and classroom size ($t = 1.98$) remain statistically significant positive predictors of reading ability. In addition, both the cross-level interaction between vocabulary and school size ($t = 2.19$) and the impact of school size alone ($t = -2.34$) are also statistically significant predictors of the reading score. The statistically significant interaction term indicates that the impact of student vocabulary score on reading achievement is dependent to some degree on the size of the school, so that the main effects for school and vocabulary cannot be interpreted in isolation, but must be considered in light of one another. The interested reader is referred to Aiken and West (1991) for more detail regarding the interpretation of interactions in regression.

The R^2 for Model4.3 appears below:

$$R_1^2 = 1 - \frac{\sigma_{M1}^2 + \tau_{M1}^2}{\sigma_{M0}^2 + \tau_{M0}^2} = 1 - \frac{3.69816 + .08856}{4.847 + .2727} = 1 - \frac{3.78672}{5.1197} = 1 - .73964 = .26.$$

By including the interaction of classroom and school size, we finish with a model that explains approximately 26% of variance in the outcome. This value is extremely similar to the portion of variance explained by the model without the interaction, further suggesting that its inclusion may not contribute much to the understanding of reading test scores.

Finally, we can determine whether or not the model including the interaction provides better fit to the data than did Model4.1, with no interaction. Once again, we will make this decision based upon the AIC/BIC values and the chi-square deviance test given by the anova() command. Given that the AIC and BIC are so similar, we can focus our interpretation on the chi-square test of deviance. The chi-square test of deviance is significant ($\chi^2 = 4.81, p < .05$) we would conclude that including the interaction between vocabulary score and school size does yield a better fitting model.

```
anova(Model4.1, Model4.2)
refitting model(s) with ML (instead of REML)
Data: Achieve
Models:
Model4.1: geread ~ gevocab + clenroll + cenroll + (1 | school/
class)
Model4.2: geread ~ gevocab + clenroll + cenroll + gevocab *
cenroll + (1 |
Model4.2:      school/class)
         Df    AIC    BIC logLik deviance  Chisq Chi Df
Pr(>Chisq)
Model4.1  7 43101 43152 -21544    43087
```

```
Model4.2  8 43099 43157 -21541    43083 4.8105       1
0.02829 *
---
Signif. codes:  0 `***' 0.001 `**' 0.01 `*' 0.05 `.' 0.1 ` ' 1
```

Defining Simple Models with More than Three Levels in the lme4 Package

To this point in the chapter, we have discussed the use of R for fitting multilevel models with three levels of data structure. In some cases, however, we may wish to fit multilevel models including more than three levels. The lmer function in R can be used to fit such higher-level models in much the same way as we have seen up to this point. As a simple example of such higher-order models, we will again fit a null model predicting reading achievement, this time incorporating four levels of data: students nested within classrooms nested within schools nested within school corporations (sometimes termed districts). In order to represent the three higher levels of influence, the random effects will now be represented as (1|corp/school/class) in Model4.3, below. In addition to fitting the model and obtaining a summary of results, we will also request bootstrapped 95% confidence intervals for the model parameters using the confint() function.

```
Model4.3 <- lmer(geread~1 + (1|corp/school/class),
data=Achieve)
summary(Model4.3)
Linear mixed model fit by REML [`lmerMod']
Formula: geread ~ 1 + (1 | corp/school/class)
   Data: Achieve

REML criterion at convergence: 46103.2

Scaled residuals:
    Min      1Q  Median      3Q     Max
-2.2995 -0.6305 -0.2131  0.3029  3.9448

Random effects:
 Groups               Name         Variance Std.Dev.
 class:(school:corp)  (Intercept)  0.27539  0.5248
 school:corp          (Intercept)  0.08748  0.2958
 corp                 (Intercept)  0.17726  0.4210
 Residual                          4.84699  2.2016
Number of obs: 10320, groups:  class:(school:corp), 568;
school:corp, 160; corp, 59
```

```
Fixed effects:
            Estimate Std. Error t value
(Intercept)  4.32583     0.07198    60.1

confint(Model4.3, method=c("profile"))
.sig01       0.4544043 0.5983855
.sig02       0.1675851 0.4083221
.sig03       0.3147483 0.5430904
.sigma       2.1710534 2.2328418
(Intercept)  4.1838496 4.4679558
```

In order to ensure that the dataset is being read by R as one thinks it should be, we can first find the summary of the sample sizes for the different data levels which occurs toward the bottom of the printout. There were 10,320 students (groups) nested within 568 classrooms (within schools within corporations) nested within 160 schools (within corporations) nested within 59 school corporations. This matches what we know about the data; therefore, we can proceed with the interpretation of the results. Given that this is a null model with no fixed predictors, our primary focus is on the intercept estimates for the random effects and their associated confidence intervals. We can see from the results confidence interval results below that each level of the data yielded intercepts that were statistically significantly different from 0 (given that 0 does not appear in any of the confidence intervals), indicating that mean reading achievement scores differed among the classrooms, the schools, and the school corporations.

Random Coefficients Models with Three or More Levels in the lme4 Package

In Chapter 2 we discussed the random coefficients multilevel model, in which the impact of one or more fixed effects is allowed to vary across the levels of a random effect. Thus, for example, we could assess whether the relationship of vocabulary test score on reading achievement differs by school. In Chapter 3 we learned how to fit such random coefficient models using lmer. Given the relative similarity in syntax for fitting two-level and three-level models, as might be expected, the definition of random coefficient models in the three-level context with lmer is very much like that for two-level models. As an example, let's consider a model in which we are interested in determining whether mean reading scores differ between males and females, while accounting for the relationship between vocabulary and reading. Furthermore, we believe that the relationship of gender to reading may differ across schools and across classrooms, thus leading to a model where the gender coefficient is allowed to vary across both random effects in our three-level model. Below is the lmer command sequence for fitting this model.

```
Model4.4 <- lmer(geread~gevocab+gender +(gender|school/class),
data=Achieve)

summary(Model4.4)
Linear mixed model fit by REML ['lmerMod']
Formula: geread ~ gevocab + gender + (gender | school/class)
   Data: Achieve

REML criterion at convergence: 43107.8

Scaled residuals:
    Min      1Q  Median      3Q     Max
-3.2043 -0.5680 -0.2069  0.3171  4.4455

Random effects:
 Groups       Name        Variance Std.Dev. Corr
 class:school (Intercept) 0.148931 0.38592
              gender      0.019807 0.14074  -0.62
 school       (Intercept) 0.033166 0.18212
              gender      0.006805 0.08249   0.58
 Residual                 3.692436 1.92157
Number of obs: 10320, groups:  class:school, 568; school, 160

Fixed effects:
             Estimate Std. Error t value
(Intercept) 2.015564   0.075629  26.651
gevocab     0.509097   0.008408  60.550
gender      0.017237   0.039243   0.439

Correlation of Fixed Effects:
        (Intr) gevocb
gevocab -0.527
gender  -0.757  0.039
```

Interpreting these results, we first see that there is not a statistically significant relationship between the fixed gender effect and reading achievement ($t = .439$). In other words, across classrooms and schools the difference in mean reading achievement for males and females is not shown to be statistically significant, when accounting for vocabulary scores. The estimate for the gender random coefficient term at the school level is approximately 0.006, and is approximately 0.02 at the classroom nested in school level. Thus, it appears that the relationship between gender and reading achievement varies more across classrooms than it does across schools, at least descriptively.

As noted above, in Model4.4, the coefficients for gender were allowed to vary randomly across both classes and schools. However, there may be some situations in which a researcher is interested in allowing the coefficient for a fixed effect to vary for only one of the random effects, such as classroom, for example. Using the syntax for Model4.5, we define the random coefficient with (gender|school/class), thus allowing both the intercept and slope to vary across both classrooms and schools. This model definition is not flexible enough to allow different random effects structures across nested levels of the data, meaning that we must allow the gender coefficient to vary across both school and classroom, if we want it to vary across the random effects at all. However, perhaps we would like to hypothesize that the relationship of gender and reading performance varies significantly across classrooms but not schools. In order to model this situation, the following syntax would be used.

```
Model4.5 <- lmer(geread~gevocab+gender + (1|school) +
(gender|class), data=Achieve)

summary(Model4.5)
Linear mixed model fit by REML ['lmerMod']
Formula: geread ~ gevocab + gender + (1 | school) + (gender |
class)
   Data: Achieve

REML criterion at convergence: 43139.9

Scaled residuals:
    Min      1Q  Median      3Q     Max
-3.0875 -0.5734 -0.2105  0.3229  4.4559

Random effects:
 Groups   Name        Variance  Std.Dev. Corr
 school   (Intercept) 9.891e-02 0.314498
 class    (Intercept) 2.566e-03 0.050653
          gender      2.307e-06 0.001519 1.00
 Residual             3.765e+00 1.940395
Number of obs: 10320, groups:  school, 160; class, 8

Fixed effects:
             Estimate Std. Error t value
(Intercept) 1.999576   0.080559  24.821
gevocab     0.513165   0.008379  61.244
gender      0.019808   0.038415   0.516

Correlation of Fixed Effects:
        (Intr) gevocb
gevocab -0.495
gender  -0.729  0.039
```

The flexibility of the lmer() definition for random effects allows for many different random effects situations to be modeled. For example, see Models4.6

and 4.7, below. Model4.6 allows gender to vary only across schools and the intercept to vary only across classrooms. Model4.7 depicts four nested levels where the intercept is allowed to vary across corporation, school, and classroom, but the gender slope is only allowed to vary across classroom.

```
Model4.6 <- lmer(geread~gevocab+gender + (-1+gender|school) +
(1|class), data=Achieve)

Model4.7 <- lmer(geread~gevocab+gender + (1|corp) + (1|school)
+ (gender|class), data=Achieve)
```

Summary

Chapter 4 is very much an extension of Chapter 3, demonstrating the use of R in fitting two-level models to include data structure at three or more levels. In practice, such complex multilevel data are relatively rare. However, as we saw in this chapter, faced with this type of data, we can use lmer to model it appropriately. Indeed, the basic framework that we employed in the two-level case works equally well for the more complex data featured in this chapter. If you have read the first four chapters, you should now feel fairly comfortable analyzing most common multilevel models with a continuous outcome variable. We will next turn our attention to the application of multilevel models to longitudinal data. Of key importance as we change directions a bit is that the core ideas that we have already learned, including the fitting of the null, random intercept, and random coefficients models, as well as inclusion of predictors at different levels of the data, do not change when we have longitudinal data. As we will see, application of multilevel models in this context is no different from that in Chapters 3 and 4. What is different is the way in which we define the levels of the data. Heretofore, level 1 has generally been associated with individual people. With longitudinal data, however, level 1 will refer to a single measurement in time, while level 2 will refer to the individual subject. By recasting longitudinal data in this manner, we make available to ourselves the flexibility and power of multilevel models.

Note

1. This is the final section of code using confint() function to boostrap confidence intervals. For full R code and output, see Chapter 3.

5

Longitudinal Data Analysis
Using Multilevel Models

To this point, we have focused on multilevel models in which a single mea-surement is made on each individual in the sample, and these individuals are in turn clustered. However, as mentioned in Chapter 2, multilevel modeling can be used in a number of different contexts, with varying data structures. In particular, this chapter will focus on using the multilevel modeling frame-work to analyze longitudinal data. Longitudinal data occur when a series of measurements are made on each individual in the sample, usually over some period of time. While there are alternatives to this temporal definition of lon-gitudinal data (e.g. measurements made at multiple locations within a plot of land), we will focus on the most common types of longitudinal data, which are time based. In this chapter, we will first demonstrate the application of tools that we have already discussed to this special case, and then briefly describe the correlation structures that are unique to longitudinal data. We will conclude the chapter by describing the advantages of using multilevel models with longitudinal data.

The Multilevel Longitudinal Framework

As with the 2- and 3-level multilevel models described in Chapters 3 and 4, longitudinal analysis in the multilevel framework involves regression-like equations at each level of the data. In the case of longitudinal models, the data structure takes the following form: repeated measurements (level 1) nested within the individual (level 2), and possibly individual nested within some higher-level cluster (e.g. school) at level 3. A simple two-level longitudi-nal model involving repeated measurements nested within individuals can be expressed as:

Level 1:

$$Y_{it} = \pi_{0i} + \pi_{1i}(T_{it}) + \pi_{2i}(X_{it}) + \varepsilon_{it} \qquad (5.1)$$

Level 2:

$$\pi_{0i} = \beta_{00} + \beta_{01}(Z_i) + r_{0i}$$
$$\pi_{1i} = \beta_{10} + r_{1i}$$
$$\pi_{2i} = \beta_{20} + r_{2i}$$

where: Y_{it} is the outcome variable for individual i at time t; π_{it} are the level-1 regression coefficients; β_{it} are the level-2 regression coefficients; ε_{it} is the level-1 error, r_{it} are the level-2 random effects; T_{it} is a dedicated time predictor variable; X_{it} is a time-varying predictor variable, and Z_i is a time invariant predictor. Thus, as can be seen in Equation (5.1), although there is new notation to define specific longitudinal elements, the basic framework for the multilevel model is essentially the same as we saw for the two-level model in Chapter 3. The primary difference is that now we have three different types of predictors: a time predictor, time-varying predictors, and time-invariant predictors. Given their unique role in longitudinal modeling, it is worth spending just a bit of time defining each of these predictor types.

Of the three types of predictors that are possible in longitudinal models, a dedicated time variable is the only one that is necessary in order to make the multilevel model longitudinal. This time predictor, which is literally an index of the time point at which a particular measurement was made, can be very flexible with time measured in fixed intervals or in waves. If time is measured in waves, they can be waves of varying length from person to person, or they can be measured on a continuum. It is important to note that when working with time as a variable it is often worthwhile to rescale the time variable so that the first measurement occasion is the zero point, thereby giving the intercept the interpretation of baseline, or initial status on the dependent variable.

The other two types of predictors (time varying and time invariant), differ in terms of how they are measured. A predictor is time varying when it is measured at multiple points in time, just as is the outcome variable. In the context of education, a time-varying predictor might be the number of hours in the previous 30 days a student has spent studying. This value could be recorded concurrently with the student taking the achievement test serving as the outcome variable. On the other hand, a predictor is time invariant when it is measured at only one point in time, and its value does not change across measurement occasions. An example of this type of predictor would be gender, which might be recorded at the baseline measurement occasion, and which is unlikely to change over the course of the data collection period. In the context of applying multilevel models to longitudinal data problems, time-varying predictors will appear at level 1 because they are associated with specific measurements, whereas time-invariant predictors will appear at level 2 or higher, because they are associated with the individual (or a higher data level) across all measurement conditions.

Person Period Data Structure

The first step in fitting multilevel longitudinal models with R is to make sure the data are in the proper longitudinal structure. Oftentimes such data are entered in what is called a person-level data structure. Person-level data structure includes one row for each individual in the dataset and one column for each variable or measurement on that individual. In the context of longitudinal data, this would mean that each measurement in time would have a separate column of its own. Although person-level data structure works well in many instances, in order to apply multilevel modeling techniques to longitudinal analyses, the data must be reformatted into what is called person period data structure. In this format, rather than having one row for each individual, person period data has one row for each time that each subject is measured, so that data for an individual in the sample will consist of as many rows as there were measurements made.

The data to be used in the following examples come from the realm of educational testing. Examinees were given a language assessment at six equally spaced times. As always, the data must first be read into R. In this case, the data are in the person-level data structure, in a file called Lang. This file includes the total language achievement test score measured over six different measurement occasions (the outcome variable), four language subtest scores (writing process and features, writing applications, grammar, and mechanics), and variables indicating student ID and school ID. Restructuring person-level data into person period format in R can be accomplished by creating a new data frame from the person-level data using the stack command. All time-invariant variables will need to be copied into the new data file, while time-variant variables (e.g. all test scores measured over the six measurement occasions) will need to be stacked in order to create person period format. The following R commands will read the Lang file into R and then rearrange the data into the necessary format

```
LangPP <- data.frame(ID=Lang$ID, school=Lang$school,
Process=Lang$Process,
Application=Lang$Application, Grammar=Lang$Grammar,
Mechanics=Lang$Mechanics,
stack(Lang, select=LangScore1:LangScore6))
```

This code takes all of the time-invariant variables directly from the raw person-level data, while also consolidating the repeated measurements into a single variable called values, and also creates a variable measuring time called ind. At this point, we may wish to do some recoding and renaming of variables. Renaming of variables can be accomplished via the names function, and recoding can be done via recode(var, recodes, as.factor. result, as.numeric.result=TRUE, levels).

For instance, we could rename the values variable to Language. The values variable is the seventh column, so in order to rename it we would use the following R code:

```
names(LangPP)[c(7)] <- c("Language").
```

We may also wish to recode the dedicated time variable, ind. Currently, this variable is not recorded numerically, but takes on the values "LangScore1," "LangScore2," "LangScore3," "LangScore4," "LangScore5," "LangScore6." Thus we may wish to recode the values to make a continuous numeric time predictor, as follows.

```
LangPP$Time <- recode(LangPP$ind,
'"LangScore1"=0; "LangScore2"=1; "LangScore3"=2;
"LangScore4"=3; "LangScore5"=4; "LangScore6"=5;', as.factor.
result=FALSE)
```

The option `as.factor.result=FALSE` tells R that the resulting values should be considered continuous. Thus, we have not only recoded the ind variable into a continuous time predictor, but also renamed it from ind to Time, and rescaled the variable such that the first time point is 0. As we noted earlier, when time is rescaled in this manner the intercept can be interpreted as the predicted outcome for baseline or time 0.

Fitting Longitudinal Models Using the lme4 Package

When the data have been restructured into person period format, fitting longitudinal models in a multilevel framework can be done in exactly the same manner as we saw in Chapters 3 and 4. As we noted before, the primary difference between the scenario described here and that appearing in previous chapters is that the nesting structure reflects repeated measurements for each individual. For example, using the Language data we just restructured in the previous section we would use the following syntax for a longitudinal random intercepts model predicting Language over time using the lme4 package:

```
Model5.0 <- lmer(Language~Time + (1|ID), data=LangPP)

summary(Model5.0)
Linear mixed model fit by REML ['lmerMod']
Formula: Language ~ Time + (1 | ID)
   Data: LangPP

REML criterion at convergence: 135165.6
```

```
Scaled residuals:
    Min      1Q Median    3Q    Max
-6.4040 -0.4986 0.0388 0.5669 4.8636
```

```
Random effects:
 Groups  Name         Variance Std.Dev.
 ID      (Intercept) 232.14    15.236
 Residual             56.65     7.526
Number of obs: 18228, groups: ID, 3038
```

```
Fixed effects:
            Estimate Std. Error t value
(Intercept) 197.21573  0.29356 671.80
Time          3.24619  0.03264  99.45
```

```
Correlation of Fixed Effects:
      (Intr)
Time -0.278
```

```
confint(Model5.0, method=c("profile"))¹
                 2.5 %      97.5 %
.sig01       14.842887   15.640551
.sigma        7.442335    7.611610
(Intercept) 196.640288  197.791167
Time          3.182207    3.310165
```

Given that we have devoted substantial time in Chapters 3 and 4 to interpreting multilevel model output, we will not spend a great deal of time here for that purpose, as it is essentially the same in the longitudinal context. However, it is important to point out that these results indicate a statistically significant positive relationship between time and performance on the language assessment ($t = 99.45$), such that scores increased over time. Also, the confidence intervals for the fixed and random effects all exclude 0, indicating that they are different from 0 in the population, i.e. statistically significant.

Adding predictors to the model also remains the same as in earlier examples, regardless of whether they are time varying or time invariant. For example, in order to add Grammar, which is time varying, as a predictor of total language scores over time, we would use the following:

```
Model5.1 <- lmer(Language~Time + Grammar +(1|ID), LangPP)
```

```
summary(Model5.1)
Linear mixed model fit by REML ['lmerMod']
Formula: Language ~ Time + Grammar + (1 | ID)
   Data: LangPP
```

```
REML criterion at convergence: 130021.1
```

```
Scaled residuals:
   Min        1Q Median      3Q      Max
-6.6286 -0.5260 0.0374  0.5788 4.6761

Random effects:
 Groups  Name         Variance Std.Dev.
 ID     (Intercept)   36.13     6.011
 Residual             56.64     7.526
Number of obs: 18216, groups: ID, 3036

Fixed effects:
            Estimate Std. Error t value
(Intercept) 73.703551  1.091368   67.53
Time         3.245483  0.032651   99.40
Grammar      0.630888  0.005523  114.23

Correlation of Fixed Effects:
        (Intr)  Time
Time    -0.075
Grammar -0.991  0.000
```

From these results, we see that, again, Time is positively related to scores on the language assessment, indicating that they increased over time. In addition, Grammar is also statistically significantly related to language test scores ($t = 114.23$), meaning that higher grammar test scores were associated with higher language scores.

If we wanted to allow the growth rate to vary randomly across individuals, we would use the following R command.

```
Model5.2 <- lmer(Language~Time + Grammar, (Time|ID),
data=LangPP)

summary(Model5.2)
Linear mixed model fit by REML ['lmerMod']
Formula: Language ~ Time + Grammar + (Time | ID)
  Data: LangPP

REML criterion at convergence: 128603.1

Scaled residuals:
   Min        1Q Median      3Q      Max
-5.3163 -0.5240 -0.0012 0.5363 5.0321

Random effects:
 Groups  Name         Variance Std.Dev. Corr
 ID      (Intercept)  21.577    4.645
         Time          3.215    1.793   0.03
 Residual             45.395    6.738
Number of obs: 18216, groups: ID, 3036
```

```
Fixed effects:
             Estimate Std. Error t value
(Intercept) 54.470619   1.002595   54.33
Time         3.245483   0.043740   74.20
Grammar      0.729119   0.005083  143.46

Correlation of Fixed Effects:
          (Intr)   Time
Time      -0.047
Grammar   -0.993  0.000
```

confint(Model5.2, method=c("profile"))[1]
```
                      2.5 %        97.5 %
.sig01        4.36967084      4.9175027
.sig02       -0.06472334      0.1250178
.sig03        1.70949317      1.8765765
.sigma        6.65369657      6.8231795
(Intercept)  51.93115382     57.0105666
Time          3.15974007      3.3312255
Grammar       0.71620566      0.7420327
```

In this model, the random effect for Time is assessing the extent to which growth over time differs from one person to the next. Results show that the random effect for Time is statistically significant, given that the 95% confidence interval does not include 0. Thus, we can conclude that growth rates in language scores over the six time points do differ across individuals in the sample.

We could add a third level of data structure to this model by including information regarding schools, within which examinees are nested. To fit this model we use the following code in R:

```
Model5.3 <- lmer(Language~Time + (1|school/ID), data=LangPP)

summary(Model5.3)
Linear mixed model fit by REML ['lmerMod']
Formula: Language ~ Time + (1 | school/ID)
   Data: LangPP

REML criterion at convergence: 134640.4

Scaled residuals:
  Min      1Q Median      3Q    Max
-6.4590 -0.5026 0.0400 0.5658 4.8580

Random effects:
 Groups     Name        Variance Std.Dev.
 ID:school  (Intercept) 187.18   13.681
 school     (Intercept)  69.11    8.313
 Residual                56.65    7.526
Number of obs: 18228, groups: ID:school, 3038; school, 35
```

```
Fixed effects:
            Estimate Std. Error t value
(Intercept) 197.33787    1.48044 133.30
Time          3.24619    0.03264  99.45

Correlation of Fixed Effects:
    (Intr)
Time -0.055
```

Using the anova() command, we can compare the fit of the three-level and two-level versions of this model.

```
anova(Model5.0, Model5.3)
refitting model(s) with ML (instead of REML)
Data: LangPP
Models:
Model5.0: Language ~ Time + (1 | ID)
Model5.3: Language ~ Time + (1 | school/ID)
    Df  AIC    BIC    logLik deviance Chisq Chi Df Pr(>Chisq)
Model5.0 4 135168 135199 -67580   135160
Model5.3 5 134648 134687 -67319   134638 521.99   1  < 2.2e-16
***
---
Signif. codes: 0 '***' 0.001 '**' 0.01 '*' 0.05 '.' 0.1 ' ' 1
```

Given that the AIC for Model5.3 is lower than that for Model5.0 where school is not included as a variable, and the chi-square test for deviance is significant ($\chi^2 = 521.99$, $p < .001$), we can conclude that inclusion of the school level of the data leads to better model fit.

Benefits of Using Multilevel Modeling for Longitudinal Analysis

Modeling longitudinal data in a multilevel framework has a number of advantages over more traditional methods of longitudinal analysis (e.g. ANOVA designs). For example, using a multilevel approach allows for the simultaneous modeling of both intraindividual change (how an individual changes over time), as well as interindividual change (differences in this temporal change across individuals). A particularly serious problem that afflicts many longitudinal studies is high attrition within the sample. Quite often, it is difficult for researchers to keep track of members of the sample over time, especially over a lengthy period of time. When using traditional techniques for longitudinal data analysis such as repeated measures ANOVA, only complete data cases can be analyzed. Thus, when there is a great deal of missing data, either a sophisticated missing data replacement method (e.g. multiple imputation) must be employed, or the researcher must work with a greatly reduced sample size. In contrast, multilevel models are able to use the available data from incomplete observations, thereby not reducing sample size as dramatically as do other approaches for modeling longitudinal data, nor requiring special missing data methods.

Repeated measures ANOVA is traditionally one of the most common methods for analysis of change. However, when it is used with longitudinal data, the assumptions upon which repeated measures rests may be too restrictive. In particular the assumption of sphericity (assuming equal variances of outcome variable differences) may be unreasonable given that variability can change considerably as time passes. On the other hand, analyzing longitudinal data from a multilevel modeling perspective does not require the assumption of sphericity. In addition, it also provides flexibility in model definition, thus allowing for information about the anticipated effects of time on error variability to be included in the model design. Finally, multilevel models can easily incorporate predictors from each of the data levels, thereby allowing for more complex data structures. In the context of longitudinal data, this means that it is possible to incorporate measurement occasion (level 1), individual (level 2), and cluster (level 3) characteristics. We saw an example of this type of analysis in Model5.3. On the other hand, in the context of repeated measures ANOVA or MANOVA, incorporating these various levels of the data would be much more difficult. Thus, the use of multilevel modeling in this context not only has the benefits listed above pertaining specifically to longitudinal analysis, but it brings the added capability of simultaneous analysis of multiple levels of influence.

Summary

In Chapter 5, we saw that the multilevel modeling tools we studied together in Chapters 2 through 4 can be applied in the context of longitudinal data. The key to this analysis is the treatment of each measurement in time as a level-1 data point, and the individuals on whom the measurements are made as level 2. Once this shift in thinking is made, the methodology remains very much the same as what we employed in the standard multilevel models in Chapters 3 and 4. By modeling longitudinal data in this way, we are able to incorporate a wide range of data structures, including individuals (level 2) nested within a higher level of data (level 3).

Note

1. This is the final section of code using confint() function to boostrap confidence intervals. For full R code and output, see Chapter 3.

6

Graphing Data in Multilevel Contexts

Graphing data is an important step in the analysis process. Far too often researchers skip the graphing of their data and move directly into analysis without the insights that can come from first giving the data a careful visual examination. It is certainly tempting for researchers to bypass data exploration through graphical analysis and move directly into formal statistical modeling, given that the models are generally the tools used to answer research questions. However, if proper attention is *not* paid to the graphing of data, the formal statistical analyses may be poorly informed regarding the distribution of variables, and relationships among them. For example, a model allowing only a linear relationship between a predictor and a criterion variable would be inappropriate if there is actually a nonlinear relationship between the two variables. Using graphical tools first, it would be possible for the researcher of such a case to see the nonlinearities, and appropriately account for them in the model. Perhaps one of the most eye-opening examples of the dangers in not plotting data can be found in Anscombe (1973). In this classic paper, Anscombe shows the results of four regression models that are essentially equivalent in terms of the means and standard deviations of the predictor and criterion variable, with the same correlation between the regressor and outcome variables in each dataset. However, plots of the data reveal drastically different relationships among the variables. Figure 6.1 shows these four datasets and the regression equation and the squared multiple correlation for each. First, note that the regression coefficients are identical across the models, as are the squared multiple correlation coefficients. However, the actual relationships between the independent and dependent variables are drastically different! Clearly, these data do not come from the same data-generating process. Thus, modeling the four situations in the same fashion would lead to mistaken conclusions regarding the nature of the relationships in the population. The moral of the story here is clear: plot your data!

The plotting capabilities in R are truly outstanding. It is capable of producing high-quality graphics with a great deal of flexibility. As a simple example, consider the Anscombe data from Figure 6.1. These data are actually included with R and can be loaded into a session with the following command:

```
data(anscombe)
```

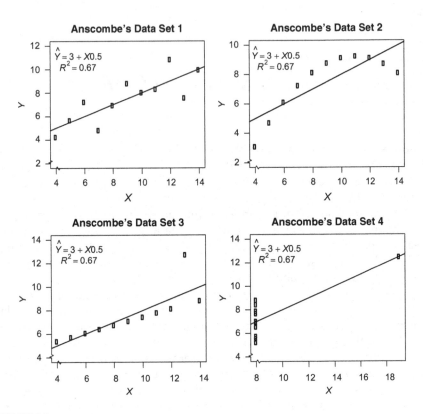

FIGURE 6.1
Plot of the Anscombe (1973) data illustration that the same set of summary statistics does not necessarily reveal the same type of information.

Examination of the data by calling upon the dataset leads to:

```
anscombe
```

	x1	x2	x3	x4	y1	y2	y3	y4
1	10	10	10	8	8.04	9.14	7.46	6.58
2	8	8	8	8	6.95	8.14	6.77	5.76
3	13	13	13	8	7.58	8.74	12.74	7.71
4	9	9	9	8	8.81	8.77	7.11	8.84
5	11	11	11	8	8.33	9.26	7.81	8.47
6	14	14	14	8	9.96	8.10	8.84	7.04
7	6	6	6	8	7.24	6.13	6.08	5.25
8	4	4	4	19	4.26	3.10	5.39	12.50
9	12	12	12	8	10.84	9.13	8.15	5.56
10	7	7	7	8	4.82	7.26	6.42	7.91
11	5	5	5	8	5.68	4.74	5.73	6.89

The way in which one can plot the data for the first dataset (i.e. x1 and y1 above), is as follows:

```
plot(anscombe$y1 ~ anscombe$x1)
```

Notice here that $y1 extracts the column labeled y1 from the data frame, as $x1 extracts the variable x1. The ~ symbol in the function call positions the data to the left as the dependent variable plotted on the ordinate (*y*-axis), whereas the value to the right is treated as an independent variable and plotted on the abscissa (*x*-axis). Alternatively, one can rearrange the terms so that the independent variable comes first, with a comma separating it from the dependent variable:

```
plot(anscombe$x1, anscombe$y1)
```

Both the former and the latter approaches lead to the same plot.

There are many options that can be called upon within the plotting framework. The `plot` function has six base parameters, with the option of calling other graphical parameters including, among others, `par` function. The `par` function has more than 70 graphical parameters that can be used to modify a basic plot. Discussing all of the available plotting parameters is beyond the scope of this chapter, however. Rather, we will discuss some of the most important parameters to consider when plotting in R.

The parameters `ylim` and `xlim` modify the starting and ending points for the *y*-and *x*-axes, respectively. For example, `ylim=c(2, 12)` will produce a plot with the *y*-axis being scaled from 2 until 12. R typically automates this process, but it can be useful for the researcher to tweak this setting, such as by setting the same axis across multiple plots. The `ylab` and `xlab` parameters are used to create labels for the *y*- and *x*-axes, respectively. For example, `ylab="Dependent Variable"` will produce a plot with the *y*-axis labeled "Dependent Variable." The `main` parameter is used for the main title of a plot. Thus, `main="Plot of Observed Values"` would produce a plot with the title "Plot of Observed Values" above the graph. There is also a sub parameter that provides a subtitle that is on the bottom of a plot, centered and below the `xlab` label. For example, `sub="Data from Study 1"` would produce such a subtitle. In some situations, it is useful to include text, equations, or a combination in a plot. Text is quite easy to include, which involves using the `text` function. For example, `text(2, 5, "Include This Text")` would place the text "Include This Text" in the plot centered at $x=2$ and $y=5$.

Equations can also be included in graphs. Doing so requires the use of `expression` within the `text` function call. The function `expression` allows for the inclusion of an unevaluated expression (i.e. it displays what is written out). The particular syntax for various mathematical expressions is available via calling `help` for the `plotmath` function (i.e. `?plotmath`). R provides a demonstration of the `plotmath` functionality via `demo(plotmath)`, which demonstrates the various mathematical expressions that can be displayed in a figure. As an example, in order to add the R^2 information to the figure in the top-left subfigure in Figure 6.1, the following command in R was used:

```
text(5.9, 9.35, expression(italic(R)^2==.67))
```

The values of 5.9 (on the *x*-axis) and 9.35 (on the *y*-axis) are simply where we thought the text looked best, and can be easily adjusted to suit the user's preferences.

Combining actual text and mathematical expressions requires using the paste function in conjunction with the expression function. For example, if we wanted to add "The Value of $R^2 = 0.67$" we would replace the previous text syntax with the following:

```
text(5.9, 9.35, expression(paste("The value of ", italic(R)^2,
" is .67", sep=""))).
```

Here, paste is used to bind together the text, which is contained within the quotes, and the mathematical expression. Note that sep="" is used so that there are no spaces added between the parts that are pasted together. Although we did not include it in our figure, the model-implied regression line can be easily included in a scatterplot. One way to do this is to go through the abline function, and include the intercept (denoted a) and the slope (denoted b). Thus, abline(a = 3, b = .5) would add a regression line to the plot that has an intercept at 3 and a slope of .5. Alternatively, to automate things a bit, using abline(lm.object) will extract the intercept and slope and include the regression line in our scatterplot.

Finally, notice in Figure 6.2 that we have broken the *y*-axis to make it very clear that the plot does not start at the origin. Whether or not this is needed may be debatable. Note that base R does not include this option, but we prefer to include it in many situations. Such a broken axis can be identified with the axis.break function from the plotrix package. The zigzag break is requested via the style="zigzag" option in axis. break, and the particular axis (1 for *y* and 2 for *x*). By default, R will set the axis to a point that is generally appropriate. However, when the origin is not shown, there is no break in the axis by default as some have argued is important.

Now, let us combine the various pieces of information that we have discussed in order to yield Figure 6.2, which was generated with the following syntax.

```
data(anscombe)

# Fit the regression model.
data.1 <- lm(y1~x1, data=anscombe)
```

```
# Scatterplot of the data.
plot(anscombe$y1 ~ anscombe$x1,
ylab=expression(italic(Y)),
ylim=c(2, 12),
xlab=expression(italic(X)),
main="Anscombe's Data Set 1")

# Add the fitted regression line.
abline(data.1)

# Add the text and expressions within the figure.
text(5.9, 9.35,
expression(paste("The value of ", italic(R)^2, " is .67",
sep="")))

text(5.9, 10.15,
expression(italic(hat(Y))==3+italic(X)*.5))


# Break the axis by adding a zigzag.
require(plotrix)
axis.break(axis=1, style="zigzag")
axis.break(axis=2, style="zigzag")
```

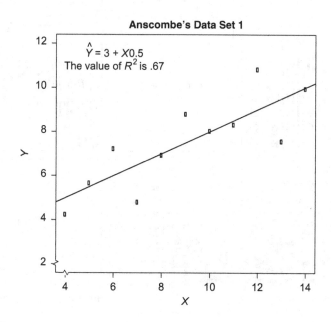

FIGURE 6.2
Plot of Anscombe's Data Set 1, with the regression line of best fit included, axis breaks, and text
denoting the regression line and value of the squared multiple correlation coefficient.

Plots for Linear Models

In order to further demonstrate graphing in R, let us recall the Cassidy GPA data from Chapter 1, in which GPA was modeled by CTA.tot and BStotal. We will now discuss some plots that are useful with single-level data, and which can be easily extended to the multilevel case, with some caveats. First, let us consider the pairs function, which plots all pairs of variables in a dataset. The resulting graph is sometimes referred to as a scatterplot matrix, because it is, in fact, a matrix of scatterplots. In the context of a multiple regression model, understanding the way in which the variables all relate to one another can be quite useful for gaining insights that might be useful as we conduct our analysis (Figure 6.3). As an example, consider the following call to pairs:

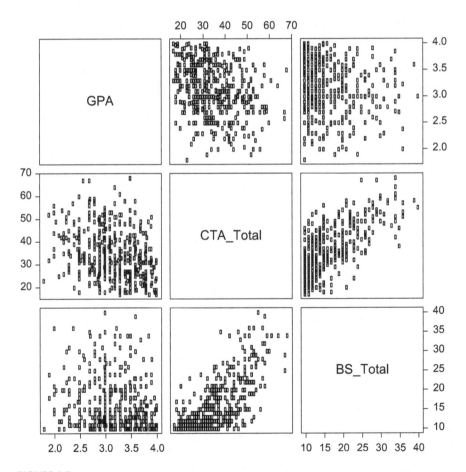

FIGURE 6.3
Pairs plot of the Cassidy GPA data. The plot shows the bivariate scatterplot for each of the three variables.

```
pairs(
cbind(GPA=Cassidy$GPA, CTA_Total=Cassidy$CTA.tot,
BS_Total=Cassidy$BStotal))
```

Given that our data contain p variables, there will be $p*(p-1)/2$ unique scatterplots (here 3). The plots below the "principal diagonal" are the same as those above the principal diagonal, the only difference being the x and y axes are reversed. Such a pairs plot allows multiple bivariate relationships to be visualized simultaneously. Of course, one can quantify the degree of linear relation with a correlation. Code to do this can be given as follows, using listwise deletion:

```
cor(na.omit(
cbind(GPA=Cassidy$GPA, CTA_Total=Cassidy$CTA.tot,
BS_Total=Cassidy$BStotal)))
```

There are other options available for dealing with missing data (see ?cor), but we have used the na.omit function wrapped around the cbind function so as to obtain a listwise deletion dataset in which the following correlation matrix is computed.

	GPA	CTA_Total	BS_Total
GPA	1.000	-0.300	-0.115
CTA_Total	-0.300	1.000	0.708
BS_Total	-0.115	0.708	1.000

Of course, when using multiple regression we must make some assumptions regarding the distribution of the model residuals, as discussed in Chapter 1. In particular, in order for the p-values and confidence intervals to be exact, we must assume that the distribution of residuals is normal. We can obtain the residuals from a model using the resid function, which is applied to a fitted lm object (e.g. GPAmodel.1 <- lm(GPA ~ CTA.tot + BStotal, data=Cass). Then, resid(GPAmodel.1)) returns the model's residuals, which can in turn be plotted in a variety of ways. One useful such plot is a histogram with an overlaid normal density curve, which can be obtained using the following R command:

```
hist(resid.1,
freq=FALSE, main="Density of Residuals for Model 1",
xlab="Residuals")
lines(density(resid.1))
```

This code first requests that a histogram be produced for the residuals. Note that freq=FALSE is used, which instructs R to make the y-axis scaled in terms of probability, not the default, which is frequency. The solid line represents the density estimate of the residuals, corresponding closely to the bars in the histogram (Figure 6.4).

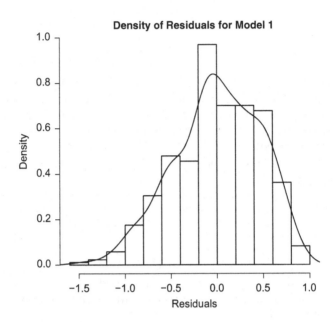

FIGURE 6.4
Histogram with overlaid density curve of the residuals from the GPA model from Chapter 1.

Rather than a histogram and overlaid density, a more straightforward way of evaluating the distribution of the errors to the normal distribution is a quantile-quantile plot (Q-Q plot). This plot includes a straight line reflecting the expected distribution of the data if in fact this is normal. In addition, the individual data points are represented as dots in the figure. When the data follow a normal distribution, the dots will fall along the straight line, or very close to it. The following code will produce the Q-Q plot that appears in Figure 6.5, based on the residuals from the GPA model.

```
qqnorm(scale(resid.1), main="Normal Quantile-Quantile Plot")
qqline(scale(resid.1))
```

Notice that above we use the scale function, which standardizes the residuals to have a mean of 0 (already done due to the regression model) and a standard deviation of 1.

We can see in Figure 6.5 that points in the Q-Q plot diverge from the line for the higher end of the distribution. This is consistent with the histogram in Figure 6.4 which shows a shorter tail on the high end as compared to the low end of the distribution. This plot, along with the histogram, reveals that the model residuals do diverge from a perfectly normal distribution. Of course, the degree of non-normality can be quantified (e.g. with skewness and kurtosis measures), but for this chapter we are most interested in visualizing the data, and especially in looking for gross violations of assumptions.

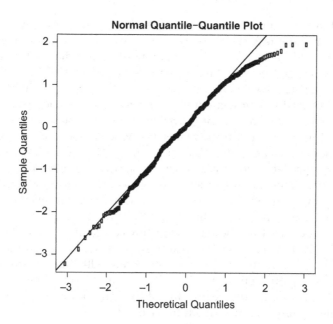

FIGURE 6.5

Plot comparing the observed quantiles of the residuals to the theoretical quantiles from the standard normal distribution. When points fall along the line (which has slope 1 and goes through the origin), then the sample quantiles follow a normal distribution.

We do need to note here that there is a grey area between an assumption being satisfied, and there being a "gross violation." At this point in the discussion, we leave the interpretation to the reader and provide information on how such visualizations can be made.

Plotting Nested Data

Heretofore, we have illustrated some basic plotting capabilities of R for linear models with only a single level. It should be noted that these tools are also potentially useful in the multilevel context, though they are not specific to it. Next, we move on to graphical tools more specifically useful with multilevel data. Multilevel models are often applied to relatively large, indeed sometimes huge, datasets. Such datasets provide a richness that cannot be realized in studies with small samples. However, a complication that often arises from this vastness of multilevel data is the difficulty of creating plots that can summarize large amounts of information and thereby provide insights into the nature of relationships among the variables. For example, the Prime Time school dataset contains more than 10,000 third-grade students. A single plot with all

10,000 individuals can be overwhelming, and fairly uninformative. Including a nesting structure (e.g. school corporations) can lead to many "corporation specific" plots (the data contain 60 corporations). Thus, the plotting of nested data necessarily carries with it more nuances (i.e. difficulties) than is typically the case with single-level data. Graphing such data and ignoring the nesting structure can lead to aggregation bias, and misinterpretation of results. For example, two variables may be negatively related within nested structures (e.g. classroom) but, overall, positively related (e.g. when ignoring classrooms). This is sometimes known as Simpson's Paradox. In addition, sometimes the nested structure of the data does not change the sign of the relationship between variables, but rather the estimate of the relationship can be strengthened or suppressed by the nested data structure, in what has been called the reversal paradox (Tu, Gunnell, and Gilthorpe, 2008). Thus, when plotting multilevel data, the nested structure should be explicitly considered. If not, at minimum a realization is necessary that the relationships in the unstructured data may differ when the nesting structure is considered. Although dealing with the complexities in plotting nested data can be at times vexing, the richness that nested data provide far outweighs any of the complications that arise when appropriately gaining an understanding from it.

Using the Lattice Package

The lattice package in R provides the researcher with several powerful tools for plotting nested data (Sarkar, 2008). One very useful function in the lattice package is dotplot. One way to use this function is to plot a variable of interest on the x-axis with a grouping variable (e.g. classrooms) on the y-axis. In the Prime Time data, there is an identifier for each corporation and for each school within corporation, as well as for each individual student. Classrooms within a school are simply numbered 1–n_j, where n_j is the number of classrooms in the school. This type of structure is typical of large multilevel datasets. Suppose that we were interested in how reading achievement (geread) is distributed within and between classrooms. To begin, let's take a single school (number 767) in a single corporation (number 940). This is the first school represented in the dataset and we are using it solely for illustrative purposes. Within this school, there are four classrooms. In order to compare the distributions of geread scores between classrooms within the school, we can use the dotplot function from the lattice package as follows:

```
dotplot(
class ~ geread,
data=Achieve.940.767, jitter.y = TRUE, ylab="Classroom",
```

```
main="Dotplot of \'geread\' for Classrooms in School 767,
Which is Within Corporation 940")
```

There are several programming points to notice here. First, we created a new dataset containing data for just this individual school, using the following code:

```
Achieve.940.767 <- Achieve[Achieve$corp==940 &
Achieve$school==767, ]
```

This R command literally identifies the rows in which the corp is equal to 940 (equality checks require two equal signs) and in which the school is equal to 767. The class ~ geread part of the code instructs the function to plot geread for each of the classrooms. The jitter.y parameter is used, which will jitter, or slightly shift around, overlapping data points in the graph. For example, if multiple students in the same classroom have the same score for geread, using jitter will shift those points on the *y*-axis to make clear that there are multiple values at the same *x* value. Finally, as before, labels can be specified. Note that the use of \'geread\' in the main title is to put geread in a single quote in the title. Calling the dotplot function using these R commands yields Figure 6.6.

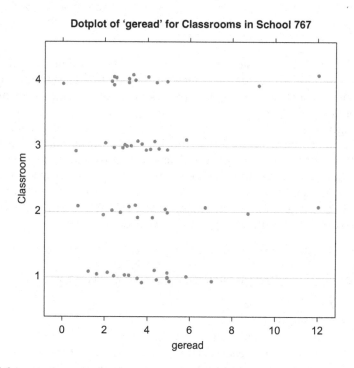

FIGURE 6.6
Dotplot of classrooms in school 767 (within corporation 940) for geread.

This dotplot shows the dispersion of geread for the classrooms in school 767. From this plot, we can see that students in each of the four classrooms had generally similar reading achievement scores. However, it is also clear that classrooms 2 and 4 each have multiple students with outlying scores that are higher than those of other individuals within the school. We hope that it is clear how a researcher could make use of this type of plot when examining the distributions of scores for individuals at a lower level of data (e.g. students) nested within a higher level (e.g. classrooms or schools).

Because the classrooms within a school are arbitrarily numbered, we can alter the order in which they appear in the graph in order to make the display more meaningful. Note that if the order of the classrooms had not been arbitrary in nature (e.g. honors classes were numbered 1 and 2), then we would need to be very careful about changing the ordering. However, in this case, no such concerns are necessary. In particular, the use of the reorder function on the left-hand side of the ~ symbol will reorder the classrooms in ascending order of the variable of interest (here, geread) in terms of the mean. Thus, we can modify Figure 6.6 in order to have the classes placed in descending order by the mean of geread.

```
dotplot(
reorder(class, geread) ~ geread,
data=Achieve.940.767, jitter.y = TRUE, ylab="Classroom",
main="Dotplot of \'geread\' for Classrooms in School 767,
Which is Within Corporation 940")
```

From Figure 6.7, it is easier to see the within-class and between-class variability for school 767. Visually, at least, it is clear that classroom 3 is more homogeneous (smaller variance) and lower performing (smaller mean) than classrooms 2 and 4.

Although dotplots such as those in Figures 6.6 and 6.7 are useful, creating one for each school would yield so much visual information that it may be difficult to draw any meaningful conclusions regarding our data. Therefore, suppose that we ignored the classrooms and schools, and instead focused on the highest level of data: corporation. Using what we have already learned, it is possible to create dotplots of student achievement for each of the corporations. To do so, we would use the following code:

```
dotplot(reorder(corp, geread) ~ geread, data=Achieve, jitter.y
= TRUE,
ylab="Classroom", main="Dotplot of \'geread\' for All
Corporations")
```

and produce Figure 6.8.

The resulting dotplots, which appear in Figure 6.8, demonstrate that with so many students within each corporation the utility of the plot is, at best, very limited, even at this highest level of data structure. This is an example of what we noted earlier regarding the difficulties in plotting nested data

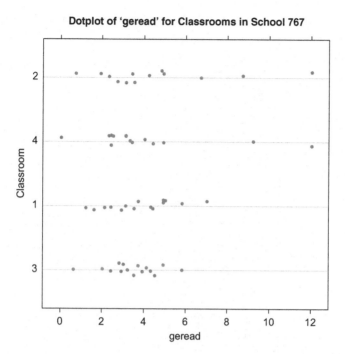

FIGURE 6.7
Dotplot of classrooms in school 767 (within corporation 940) for geread, with the classrooms ordered by the mean (lowest to highest).

due to (a) the sheer volume of the data and (b) remaining sensitive to the nesting structure of the data.

One method for visualizing such large and complex data would be to focus on a higher level of data aggregation, such as the classroom, rather than on the individual students. Recall, however, that classrooms are simply numbered from 1 to *n* within each school, and are not, therefore, given identifiers that mark them as unique across the entire dataset. Therefore, if we wish to focus on achievement at the classroom level, we must first create a unique classroom number. We can use the following R code in order to create such a unique identifier, which augments to the Achieve data with a new column of the unique classroom identifiers, which we call Classroom _ Unique.

```
Achieve <- cbind(Achieve, Classroom_Unique=paste(Achieve$corp,
Achieve$school, Achieve$class, sep=""))
```

After forming this unique identifier for the classrooms, we will then aggregate the data within the classrooms in order to obtain the mean of the variables within those classrooms. We do this by using the aggregate function:

```
Achieve.Class_Aggregated <- aggregate(Achieve,
by=list(Achieve$Classroom_Unique), FUN=mean)
```

Dotplot of 'geread' for All Corporations

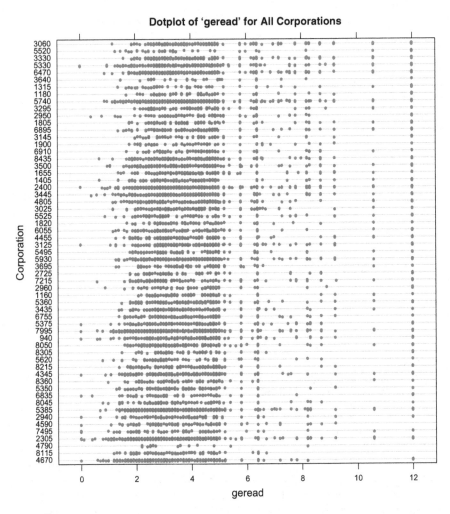

FIGURE 6.8

Dotplot of students in corporations, with the corporations ordered by the mean (lowest to highest).

This code creates a new dataset (Achieve.Class_Aggregated) that contains the aggregated classroom data. The by=list(Achieve$Classroom_Unique) part of the code instructs the function on which variable name (here Classroom_Unique) the aggregation is to be implemented. Now, with the new dataset, Achieve.Class_Aggregated, we can examine the distribution of geread means of the individual classrooms. Thus, our dataset has functionally been reduced from over 10,000 students to 568 classrooms. We create the dotplot with the following command:

```
dotplot(reorder(corp, geread) ~ geread, data=Achieve.
Class_Aggregated,
```

```
jitter.y = TRUE,
ylab="Corporation", main="Dotplot of Classroom Mean \'geread\'
Within the Corporations")
```

Of course, we still know the nesting structure of the classrooms within the schools and the schools within the corporations. We are aggregating here for purposes of plotting, but not modeling the data. We want to remind readers of the potential dangers of aggregation bias discussed earlier in this chapter. With this caveat in mind, consider Figure 6.9, which shows that within the corporations, classrooms do vary in terms of their mean level of achievement (i.e. the within line/corporation spread) as well as between corporations (i.e. the change in the lines).

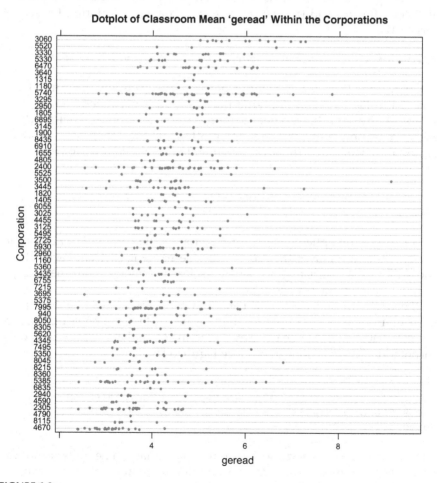

FIGURE 6.9
Dotplot of geread for corporations, with the corporations ordered by the mean (lowest to highest) of the classroom aggregated data. The dots in the plot are the mean of the classrooms scores within each of the corporations.

We can also use dotplots to gain insights into reading performance within specific school corporations. But, again, this would yield a unique plot, such as the one above, for each corporation. Such a graph may be useful when interest concerns a specific corporation, or when one wants to assess variability in specific corporations.

We hope to have demonstrated that dotplots can be useful tools for gaining an understanding of the variability that exists, or that does not exist, in the variable(s) of interest. Of course, looking only at a single variable can be limiting. Another particularly useful function for multilevel data that can be found in the lattice package is xyplot. This function creates a graph very similar to a scatterplot matrix for a pair of variables, but accounting for the nesting structure in the data. For example, the following code produces an xyplot for geread (*y*-axis) by gevocab (*x*-axis), accounting for school corporation.

```
xyplot(geread ~ gevocab | corp, data = Achieve)
```

Notice that here the | symbol defines the grouping/nesting structure, with the ~ symbol implying that geread is predicted/modeled by gevocab. By default, the specific names of the grouping structure (here the corporation numbers) are not plotted on the "strip." To produce Figure 6.10, we added the strip argument with the following options to the above code

```
strip=strip.custom(strip.names=FALSE, strip.levels=c(FALSE,
TRUE))
```

Our use of the optional strip argument adds the corporation number to the graph, and removes the "corp" variable name itself from each "strip" above all of the bivariate plots, which itself was removed with the strip. names=FALSE subcommand.

Of course, any sort of specific conditioning of interest can be done for a particular graph. For example, we might want to plot the schools in, say, corporation 940, which can be done by extracting from the Achieve data only corporation 940, as is done to produce Figure 6.11.

```
xyplot(geread ~ gevocab | school, data =
Achieve[Achieve$corp==940,], strip=strip.custom(strip.
names=FALSE, strip.levels=c(FALSE, TRUE)), main="Schools in
Corporation 940")
```

We have now discussed two lattice functions that can be quite useful for visualizing grouped/nested data. An additional plotting strategy involves assessment of the residuals from a fitted model, as doing so can help discern when there are violations of assumptions, much as we saw earlier in this chapter when discussing single-level regression models. Because residuals

Students in Each Corporation

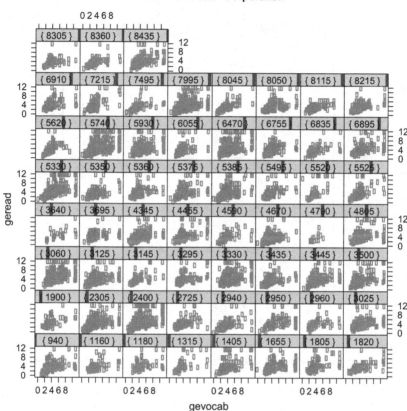

FIGURE 6.10

Xyplot from of geread (*y*-axis) as a function of gevocab (*x*-axis) by corporation.

are assumed to be uncorrelated with any of the grouping structures in the model, they can be plotted using the R functions that we discussed earlier for single-level data. For example, to create a histogram with a density line plot of the residuals, we first standardized the residuals and used code as we did before. Figure 6.12 is produced with the following syntax:

```
hist(scale(resid(Model3.1)),
freq=FALSE, ylim=c(0, .7), xlim=c(-4, 5),
main="Histogram of Standardized Residuals from Model 3.1",
xlab="Standardized Residuals")
lines(density(scale(resid(Model3.1))))
box()
```

The only differences in the way that we plotted residuals with hist earlier in the chapter are purely cosmetic in nature. In particular, here we used the

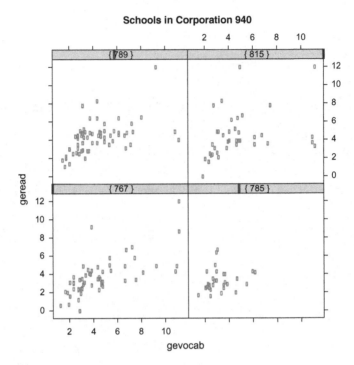

FIGURE 6.11
Xyplot from of geread (*y*-axis) as a function of gevocab (*x*-axis) by school within corporation 940.

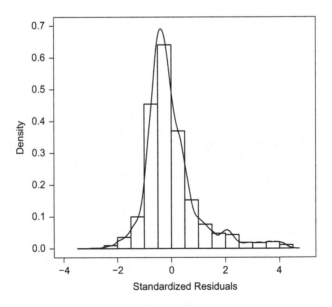

FIGURE 6.12
Histogram and density plot for standardized residuals from the Model 3.1.

box function to draw a box around the plot and specified the limits of the
y- and *x*-axis.

Alternatively, a Q-Q plot can be used to evaluate the assumption of normality, as described earlier in the chapter (Figure 6.13). The code to do such a plot is:

```
qqnorm(scale(resid(model3.1)))
qqline(scale(resid(model3.1)))
```

Clearly, the Q-Q plot (and the associated histogram) illustrate that there are
issues on the high end of the distribution of residuals. The issue, as it turns
out, is not so uncommon in educational research: ceiling effects. In particular,
an examination of the previous plots we have created reveals that there are is
non-trivial number of students who achieved the maximum score on geread.
The multilevel model assumes that the distribution of residuals follows a normal distribution. However, when a maximum value is reached, it is necessarily the case that the residuals will not be normally distributed because, as in
this case, a fairly large number of individuals have the same residual value.

The plotting capabilities available in R are impressive. Unfortunately, we
were only able to highlight a few of the most useful plotting functions in this
chapter. For the most part, the way to summarize the graphical ability of R
and the available packages is, "if you can envision it, you can implement it."
There are many great resources available for graphics in R, and an internet
search will turn up many great online resources that are freely available.

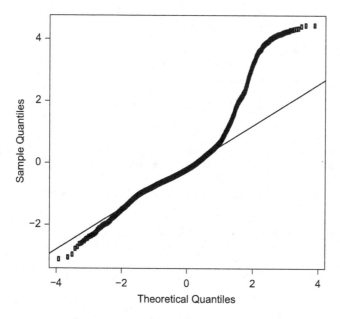

FIGURE 6.13
Q-Q plot of the standardized residuals from Model 3.1.

Plotting Model Results Using the Effects Package

As a part of our interpretation of statistical modeling results, we may find it helpful to examine visual representations of the relationships between the independent and dependent variables of interest. The effects R library will be particularly helpful in this regard. This package can be used with the results of any of several statistical modeling libraries, including lm and lme4. In order to see how we can use effects, let's consider the data that we used in Chapter 3, here called prime _ time. In this example, we will fit a model in which the dependent variable is the reading score, geread, and the predictors are measures of verbal (npaverb) and nonverbal (npanverb) reasoning skills. The first model that we fit appears below, followed by the summary output.

```
Model6.1 <- lmer(geread~npaverb  + (1|school),
data=prime_time)
summary(Model6.1)
Linear mixed model fit by REML. t-tests use Satterthwaite's
method ['lmerModLmerTest']
Formula: geread ~ npaverb + (1 | school)
   Data: prime_time

REML criterion at convergence: 45374.6

Scaled residuals:
    Min       1Q  Median       3Q      Max
-2.1620  -0.6063  -0.1989   0.3045   4.8067

Random effects:
 Groups    Name         Variance Std.Dev.
 school    (Intercept)  0.1063   0.326
 Residual               3.8986   1.974
Number of obs: 10765, groups:   school, 163

Fixed effects:
             Estimate Std. Error        df t value Pr(>|t|)
(Intercept) 1.952e+00  5.146e-02 8.143e+02   37.94   <2e-16
***
npaverb     4.390e-02  7.372e-04 9.585e+03   59.56   <2e-16
***
---
Signif. codes:  0 '***' 0.001 '**' 0.01 '*' 0.05 '.' 0.1 ' ' 1

Correlation of Fixed Effects:
        (Intr)
npaverb -0.775
```

From these results, we can see that the fixed effect npaverb has a statistically significant positive relationship with geread. Using the effects package, we can visualize this relationship in the form of the line of best fit, with an accompanying confidence interval. In order to obtain this graph, we will use the command sequence below.

```
library(effects)
plot(predictorEffects(Model6.1))
```

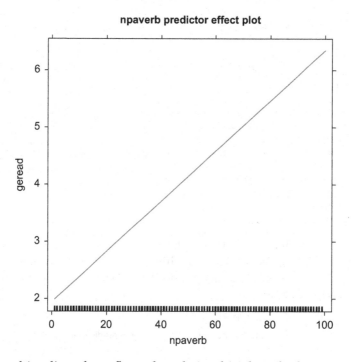

The resulting line plot reflects the relationship described numerically by the npaverb slope of 0.0439. In order to plot the relationship, we simply employ the `predictorEffects` command in R.

Multiple dependent variables can be plotted at once using this basic command structure. In Model6.2, we include school socioeconomic status (SES) in our model.

```
Model6.2 <- lmer(geread~npaverb + ses +(1|school),
data=prime_time)
summary(Model6.2)

Linear mixed model fit by REML. t-tests use Satterthwaite's
method ['lmerModLmerTest']
Formula: geread ~ npaverb + ses + (1 | school)
   Data: prime_time

REML criterion at convergence: 45376.9
```

```
Scaled residuals:
    Min      1Q  Median      3Q     Max
-2.1458 -0.6085 -0.1988  0.3057  4.7966

Random effects:
 Groups    Name         Variance Std.Dev.
 school    (Intercept)  0.09759  0.3124
 Residual               3.89872  1.9745
Number of obs: 10765, groups:  school, 163

Fixed effects:
              Estimate Std. Error        df t value Pr(>|t|)
(Intercept) 1.662e+00 1.077e-01 1.801e+02  15.437  < 2e-16
***
npaverb     4.356e-02 7.467e-04 1.055e+04  58.345  < 2e-16 ***
ses         4.290e-03 1.413e-03 1.709e+02   3.036  0.00277 **
---
Signif. codes:  0 '***' 0.001 '**' 0.01 '*' 0.05 '.' 0.1 ' ' 1

Correlation of Fixed Effects:
        (Intr) npavrb
npaverb -0.217
ses     -0.881 -0.168
```

From these results, we see that higher levels of school SES are associated with higher reading test scores, as are higher scores on the verbal reasoning test. We can plot these relationships and the associated confidence intervals simultaneously as below.

```
plot(predictorEffects(Model6.2))
```

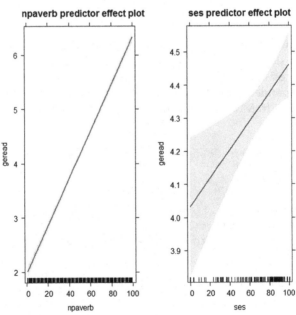

It is also possible to plot only one of these relationships per graph. We also added a more descriptive title for each of the following plots, using the `main` subcommand.

```
plot(predictorEffects(Model6.2, ~npaverb), main="Reading by
Verbal Reasoning")
```

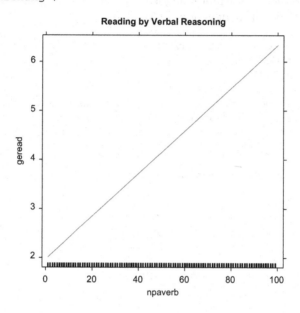

```
plot(predictorEffects(Model6.2, ~ses), main="Reading by School
SES")
```

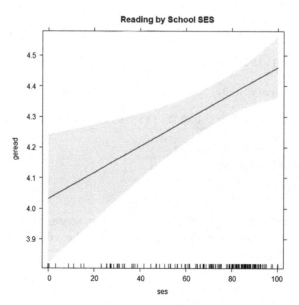

One very useful aspect of these plots is the inclusion of the 95% confidence region around the line. From this, we can see that our confidence regarding the actual nature of the relationships in the population is much greater for verbal reasoning than for school SES. Of course, given the much larger level-1 sample size, this makes perfect sense.

It is also possible to include categorical independent variables along with their plots, as with student gender in Model6.3.

```
Model6.3 <- lmer(geread~npaverb + gender +(1|school),
data=prime_time)
summary(Model6.3)

Linear mixed model fit by REML. t-tests use Satterthwaite's
method ['lmerModLmerTest']
Formula: geread ~ npaverb + gender + (1 | school)
   Data: prime_time

REML criterion at convergence: 45200.6

Scaled residuals:
    Min      1Q  Median      3Q     Max
-2.1691 -0.6094 -0.1993  0.3040  4.8117

Random effects:
 Groups    Name        Variance Std.Dev.
 school    (Intercept) 0.106    0.3256
 Residual              3.903    1.9756
Number of obs: 10720, groups:  school, 163

Fixed effects:
              Estimate Std. Error       df t value Pr(>|t|)
(Intercept) 1.991e+00  7.783e-02 3.290e+03  25.586   <2e-16
***
npaverb     4.396e-02  7.387e-04 9.533e+03  59.506   <2e-16
***
gender     -2.765e-02  3.835e-02 1.065e+04  -0.721    0.471
---
Signif. codes:  0 '***' 0.001 '**' 0.01 '*' 0.05 '.' 0.1 ' ' 1

Correlation of Fixed Effects:
        (Intr) npavrb
npaverb -0.528
gender  -0.749  0.020
```

Gender is not statistically related to reading score, as reflected in the *p*-value of 0.471, and in the graph below. Note the very wide confidence region around the line.

```
plot(predictorEffects(Model6.3, ~gender))
```

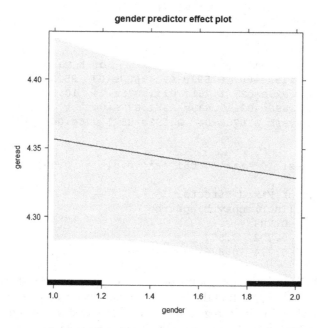

Finally, we can also use the `effects` package to graphically probe interactions among independent variables. For example, consider Model6.4, in which the variables npaverb and npanverb (nonverbal reasoning) are used as predictors of reading score. In addition, the interaction of these two variables is also included in the model.

```
Model6.4 <- lmer(geread~npaverb + npanverb + npaverb*npanverb
+(1|school), data=prime_time)
summary(Model6.4)

Linear mixed model fit by REML. t-tests use Satterthwaite's
method ['lmerModLmerTest']
Formula: geread ~ npaverb + npanverb + npaverb * npanverb + (1
 | school)
   Data: prime_time

REML criterion at convergence: 44948.9

Scaled residuals:
    Min       1Q   Median       3Q      Max
-2.2001  -0.6070  -0.1904   0.3154   4.8473

Random effects:
 Groups    Name          Variance Std.Dev.
```

```
 school    (Intercept) 0.09077  0.3013
 Residual              3.74551  1.9353
Number of obs: 10762, groups:   school, 163

Fixed effects:
           Estimate Std. Error        df t value Pr(>|t|)
(Intercept) 2.192e+00  8.592e-02 4.637e+03  25.516   <2e-16 ***
npaverb    1.923e-02  1.761e-03 1.075e+04  10.920   <2e-16 ***
npanverb   2.896e-03  1.639e-03 1.075e+04   1.766   0.0774 .
npaverb:npanverb 2.678e-04  2.692e-05 1.076e+04   9.950    <2e-
16 ***
---
Signif. codes:  0 '***' 0.001 '**' 0.01 '*' 0.05 '.' 0.1 ' ' 1

Correlation of Fixed Effects:
           (Intr) npavrb npnvrb
npaverb     -0.783
npanverb    -0.792  0.577
npvrb:npnvr  0.774 -0.866 -0.844
```

These results indicate that nonverbal reasoning is not statistically related to reading test score, but that there is an interaction between verbal and nonverbal reasoning. In order to gain a better understanding as to the nature of this interaction, we can plot it using the `effects` package.

```
plot(predictorEffects(Model6.4, ~npaverb*npanverb))
```

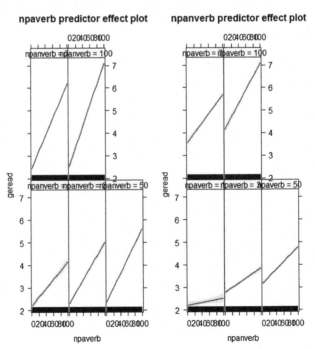

The interaction effect is presented in two plots, one for each variable. The set of line plots on the right focuses on the relationship between npaverb and geread, at different levels of npanverb. Conversely, the set of figures on the left switches the focus to the relationship of npanverb and reading score at different levels of npaverb. Let's focus on the right-side plot. We see that R has arbitrarily plotted the relationship between npanverb and geread at 5 levels of npaverb, 1, 30, 50, 70, and 100. The relationship between npanverb and geread is very weak when npaverb=1, and increases in value (i.e. the line becomes steeper) as the value of npaverb increases in value. Therefore, we would conclude that nonverbal reasoning is more strongly associated with reading test score for individuals who exhibit a higher level of verbal reasoning. For a discussion as to the details of how these calculations are made, the reader is referred to the effects package documentation (https://cran.r-project.org/web/packages/effects/effects.pdf).

It is also possible to plot three-way interactions using the effects package. Model6.5 includes a measure of memory, in addition to the verbal and nonverbal reasoning scores. The model summary information demonstrates that there is a statistically significant three-way interaction among the independent variables.

```
Model6.5 <- lmer(geread~npaverb * npanverb * npamem
+(1|school), data=prime_time)
summary(Model6.5)

Linear mixed model fit by REML. t-tests use Satterthwaite's
method ['lmerModLmerTest']
Formula: geread ~ npaverb * npanverb * npamem + (1 | school)
   Data: prime_time

REML criterion at convergence: 44987.7

Scaled residuals:
    Min      1Q  Median      3Q     Max
-2.3007 -0.6041 -0.1930  0.3169  4.8402

Random effects:
 Groups   Name        Variance Std.Dev.
 school   (Intercept) 0.08988  0.2998
 Residual             3.73981  1.9339
Number of obs: 10758, groups:  school, 163

Fixed effects:
                          Estimate Std. Error          df t
value Pr(>|t|)
(Intercept)              2.028e+00  1.594e-01  9.900e+03
12.722  < 2e-16 ***
```

```
npaverb                              2.162e-02   3.665e-03   1.072e+04
5.899 3.76e-09 ***
npanverb                             1.004e-02   3.577e-03   1.071e+04
2.808     0.0050 **
npamem                               2.796e-03   2.959e-03   1.071e+04
0.945     0.3446
npaverb:npanverb                     1.235e-04   6.032e-05   1.069e+04
2.048     0.0406 *
npaverb:npamem                      -3.065e-05   6.288e-05   1.070e+04
-0.487    0.6259
npanverb:npamem                     -1.157e-04   5.986e-05   1.070e+04
-1.933    0.0533 .
npaverb:npanverb:npamem  2.169e-06   9.746e-07   1.069e+04
2.225     0.0261 *
---
Signif. codes:  0 '***' 0.001 '**' 0.01 '*' 0.05 '.' 0.1 ' ' 1

Correlation of Fixed Effects:
            (Intr) npavrb npnvrb npamem npvrb:npn npvrb:npm
npnvr:
npaverb     -0.783

npanverb    -0.778  0.509
npamem      -0.837  0.652  0.643
npvrb:npnvr  0.775 -0.834 -0.832 -0.629
npavrb:npmm  0.708 -0.869 -0.449 -0.811  0.712
npnvrb:npmm  0.720 -0.464 -0.878 -0.816  0.722      0.563
npvrb:npnv: -0.730  0.761  0.749  0.802 -0.883     -0.857
-0.845
```

The interaction can be plotted using effects.

```
plot(predictorEffects(Model6.5, ~npaverb*npanverb*npamem))
```

For the two-way interaction that was featured in Model6.4, we saw that the relationship between npanverb and geread was stronger for higher levels of npaverb. If we focus on the relationship between npanverb and geread in the three-way interaction, we see that this pattern continues to hold, and that it is amplified somewhat for larger values of npamem. In other words, the relationship between nonverbal reasoning and reading test score is stronger for larger values of verbal reasoning, and stronger still for the combination of higher verbal reasoning and higher memory scores.

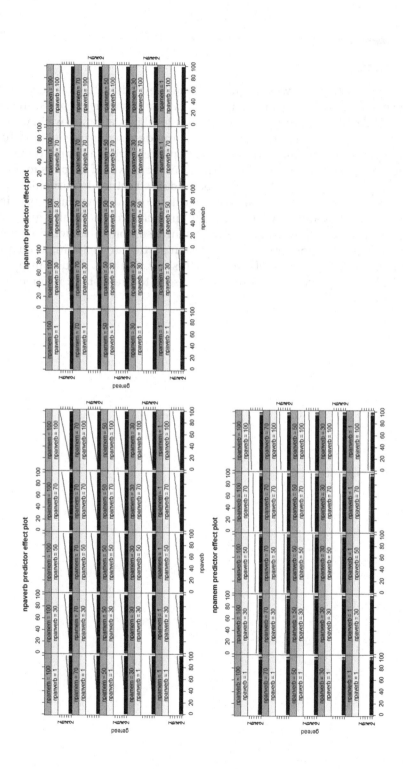

Summary

The focus of Chapter 6 was on graphing multilevel data. Exploration of data using graphs is always recommended for any data analysis problem, and can be particularly useful in the context of multilevel modeling, as we have seen here. We saw how a scatterplot matrix can provide insights into relationships among variables that may not be readily apparent from a simple review of model coefficients. In addition, we learned of the power of dotplots to reveal interesting patterns at multiple levels of the data structure. In particular, with dotplots we were able to visualize mean differences among classrooms in a school, as well as among individuals within a classroom. Finally, graphical tools can also be used to assess the important assumptions underlying linear models in general, and multilevel models in particular, including normality and homogeneity of residual variance. In short, analysts should always be mindful of the power of pictures as they seek to understand relationships in their data.

7

Brief Introduction to Generalized Linear Models

Heretofore, we have focused our attention primarily on models for data in which the outcome variables are continuous in nature. Indeed, we have been even more specific and dealt almost exclusively with models resting on the assumption that the model errors are normally distributed. However, in many applications, the outcome variable of interest is categorical, rather than continuous. For example, a researcher might be interested in predicting whether or not an incoming freshman is likely to graduate from college in four years, using high-school grade point average and admissions test scores as the independent variables. Here, the outcome is the dichotomous variable graduation in four years (yes or no). Likewise, consider research conducted by a linguist who has interviewed terminally ill patients and wants to compare the number of times those patients use the word death or dying during the interviews. The number of times that each word appears, when compared to the many thousands of words contained in the interviews, is likely to be very small, if not zero for some people. Another way of considering this outcome variable is as the rate of certain target words occurring out of all of the words used by the interviewees. Again, this rate will likely be very low, so that the model errors are almost assuredly not normally distributed. Yet another example of categorical outcome variables would occur when a researcher is interested in comparing scores by treatment condition on mathematics performance outcome that are measured on a Likert scale, such as 1, 2, or 3, where higher scores indicate better performance on the mathematics task. Thus, the multilevel models that we described in Chapters 2, 3, 4, and 5 would not be applicable to these research scenarios.

In each of the previous examples, the outcome variable of interest is not measured on a continuous scale, and will almost surely not produce normally distributed model errors. As we have seen, the linear multilevel models discussed previously work under the assumption of normality of errors. As such, they will not be appropriate for situations in which these, or other types of variables that cannot be appropriately modeled with a linear model, are to be used. However, alternative models for such variables are available, thereby forcing us to consider alternative models. Taken together, these alternatives for categorical outcome variables are often referred to as generalized linear models (GLiMs). Prior to discussing the multilevel versions of these models in Chapter 8, it will behoove us to first explore some common GLiMs and their applications in the single-level context. In the next chapter, we will then expand upon our discussion here to include the multilevel variants of these models, and how to fit them in R. In the following sections of this chapter, we will focus on three broad types of GLiMs, including those for categorical outcomes (dichotomous, ordinal, and nominal), counts or rates of events that occur very infrequently, and counts or rates of events that occur somewhat more frequently. After their basic theoretical presentation, we will then describe how these single-level GLiMS can be fit using functions in R.

Logistic Regression Model for a Dichotomous Outcome Variable

As an example of a GLiM, we begin the discussion of models for dichotomous outcome data. Let's consider an example involving a sample of 20 men, 10 of whom have been diagnosed with coronary artery disease, and 10 who have not. Each of the 20 individuals was asked to walk on a treadmill until they became too fatigued to continue. The outcome variable in this study was the diagnosis and the independent variable was the time walked until fatigue; i.e. the point at which the subject requested to stop. The goal of the study was to find a model predicting coronary artery status as a function of time walked until fatigue. If an accurate predictive equation could be developed, it might be a helpful tool for physicians to use in helping to diagnose heart problems in patients. In the context of Chapter 1, we might consider applying a linear regression model to these data, as we found that this approach was useful for estimating predictive equations. However, recall that there were a number of assumptions upon which appropriate inference in the context of linear regression depends, including that the residuals would be normally distributed. Given that the outcome variable in the current problem is a dichotomy (coronary disease or not), the residuals will almost certainly not follow a normal distribution. Therefore, we need to identify an alternative approach for dealing with dichotomous outcome data such as these.

Perhaps the most commonly used model for linking a dichotomous outcome variable with one or more independent variables (either continuous or categorical) is logistic regression. The logistic regression model takes the form

$$\ln\left(\frac{p(Y=1)}{1-p(Y=1)}\right) = \beta_0 + \beta_1 x. \qquad (7.1)$$

Here, y is the outcome variable of interest, and taking the values 1 or 0, where 1 is typically the outcome of interest. (Note that these dichotomous outcomes could also be assigned other values, though 1 and 0 are probably the most commonly used in practice.) This outcome is linked to an independent variable, x, by the slope (β_1) and intercept (β_0). Indeed, the right side of this equation should look very familiar, as it is identical to the standard linear regression model. However, the left side is quite different from what we see in linear regression, due to the presence of the logistic link function, also known as the logit. Within the parentheses lie the odds that the outcome variable will take the value of 1. For our coronary artery example, 1 is the value for having coronary artery disease and 0 is the value for not having it. In order to render the relationship between this outcome and the independent variable (time walking on treadmill until fatigue) linear, we need to take the natural log of these odds. Thus, the logit link for this problem is the natural log of the odds of an individual having coronary artery disease. Interpretation of the slope and intercept in the logistic regression model are the same as interpretation in the linear regression context. A positive value of β_1 would indicate that the larger the value of x, the greater the log odds of the target outcome occurring. The parameter β_0 is the log odds of the target event occurring when the value of x is 0. Logistic regression models can be fit easily in R using the GLM function, a part of the MASS library, which is a standard package included with the basic installation of R. In the following section, we will see how to call this function and how to interpret the results we obtain from it.

The data were read into a data frame called coronary, using the methods outlined in the chapter on data management in R. The logistic regression model can then be fit in R using the following command sequence, where group refers to the outcome variable, and time is the number of seconds walked on the treadmill.

```
coronary.logistic<-glm(group~time, family=binomial)
```

Here we have created a model output object entitled coronary.logistic, which contains the parameter estimates and model fit information. The command glm indicates that we are using a GLiM, which we define within the parentheses (not to be confused with a GLM, which is fitted with lm() function in R, as we saw in Chapter 1). As with other R functions demonstrated in this book, the dependent variable appears on the left side of the ~, and the independent variable(s) appear on the right side. Finally, we indicate that this is a dichotomous logistic regression model with the family=binomial command.

We can obtain the summary from this analysis using the summary(coronary.logistic) command. When interpreting logistic regression results, it is important to know which of the two possible outcomes is being modeled by the software in the numerator of the logit. In other words, we need to know which category was defined as the target by the software so that we can properly interpret the model parameter estimates.

By default, the glm command will treat the higher value as the target. In this case 0 = healthy and 1 = disease. Therefore, the numerator of the logit will be 1, or disease. It is possible to change this so that the lower number is the target, and the interested reader can refer to help(glm) for more information in this regard. This is a very important consideration, as the results would be completely misinterpreted if R used a different specification than the user thinks was used. The results of the summary command appear below.

```
Call:
glm(formula = group ~ time, family = binomial)

Deviance Residuals:
    Min      1Q    Median      3Q      Max
-2.1387 -0.3077  0.1043  0.5708  1.5286

Coefficients:
               Estimate Std. Error z value Pr(>|z|)
(Intercept)   13.488949    5.876693   2.295   0.0217 *
coronary$time -0.016534    0.007358  -2.247   0.0246 *
---
Signif. codes: 0 '***' 0.001 '**' 0.01 '*' 0.05 '.' 0.1 ' ' 1

(Dispersion parameter for binomial family taken to be 1)

    Null deviance: 27.726  on 19  degrees of freedom
Residual deviance: 12.966  on 18  degrees of freedom
AIC: 16.966

Number of Fisher Scoring iterations: 6
```

Our initial interest is in determining whether there is a significant relationship between the independent (time) and the dependent (coronary disease status) variable. Thus, we will first take a look at the row entitled time, as it contains this information. The column labeled Estimate includes the slope and intercept values. The estimate of β_1 is −0.016534, indicating that the more time an individual could walk on the treadmill before becoming fatigued, the lower the log odds that they had coronary artery disease; i.e. the less likely they were to have heart disease. Through the simple transformation of the slope e^{β_1}, we can obtain the odds of having coronary artery disease as a function of time. For this example, $e^{-0.016534}$ is 0.984, indicating that for every additional second an individual is able to walk on the treadmill before becoming fatigued, their estimated odds of having heart disease are multiplied by 0.984. Thus, for an additional minute of walking, the odds decrease by exp(−.016534*60) = .378.

Adjacent to the coefficient column is the standard error, which measures the sampling variation in the parameter estimate. The estimate divided by the standard error yields the test statistic, which appears under the z column.

This is the test statistic for the null hypothesis that the coefficient is equal to 0. Next to z is the p-value for the test. Using standard practice, we would conclude that a p-value less than 0.05 indicates statistical significance. In addition, R provides a simple heuristic for interpreting these results based on the *. For this example, the p-value of 0.0246 for time means that there is a statistically significant relationship between time on the treadmill to fatigue and the odds of an individual having coronary artery disease. The negative sign for the estimate further tells us that more time spent on the treadmill was associated with a lower likelihood of having heart disease.

One common approach to assessing the quality of the model fit to the data is by examining the deviance values. For example, the residual deviance compares the fit between a model that is fully saturated, meaning that it perfectly fits the data, and our proposed model. Residual deviance is measured using a χ^2 statistic that compares the predicted outcome value with the actual value, for each individual in the sample. If the predictions are very far from the actual responses, this χ^2 will tend to be a large value, indicating that the model is not very accurate. In the case of the residual deviance, we know that the saturated model will always provide optimal fit to the data at hand, though in practice it may not be particularly useful for understanding the relationship between x and y in the population because it will have a separate model parameter for every cell in the contingency table relating the two variables, and thus may not generalize well. The proposed model will always be more parsimonious than the saturated (i.e. have fewer parameters), and will therefore generally be more interpretable and generalizable to other samples from the same population, assuming that it does in fact provide adequate fit to the data. With appropriately sized samples, the residual deviance can be interpreted as a true χ^2 test, and the p-value can be obtained in order to determine whether the fit of the proposed model is significantly worse than that of the saturated model. The null hypothesis for this test is that model fit is adequate; i.e. the fit of the proposed model is close to that of the saturated model. With a very small sample such as this one, this approximation to the χ^2 distribution does not hold (Agresti, 2002), and we must therefore be very careful in how we interpret the statistic. For pedagogical purposes, let's obtain the p-value for a χ^2 of 12.966 with 18 degrees of freedom. This value is 0.7936, which is larger than the α cut-off of 0.05, indicating that we cannot reject that the proposed model fits the data as well as the saturated model. Thus, we would then retain the proposed model as being sufficient for explaining the relationships between the independent and dependent variables.

The other deviance statistic for assessing fit that is provided by R is the null deviance, which tests the null hypothesis that the proposed model does not fit the data better than a model in which the average odds of having coronary artery disease are used as the predicted outcome for every time value (i.e. that x is not linear predictive of the probability of having coronary heart disease). A significant result here would suggest that the proposed model is better than no model at all. Again, however, we must interpret this test with caution when our sample size is very small, as is the case here. For this example, the p-value of the null deviance test ($\chi^2 = 27.726$ with 19 degrees of freedom) was 0.0888. As with the residual deviance test, the result is not statistically significant at $\alpha = 0.05$, suggesting that the proposed model does not provide better fit than the null model with no relationships. Of course, given the small sample size, we must interpret both hypothesis tests with some caution.

Finally, R also provides the AIC value for the model. As we have seen in previous chapters, AIC is a useful statistic for comparing the fit of different, and not necessarily nested, models with smaller values indicating better relative fit. If we wanted to assess whether including additional independent variables or interactions improved model fit, we could compare AIC values among the various models to ascertain which was optimal. For the current example, there are no other independent variables of interest. However, it is possible to obtain the AIC for the intercept-only model using the following command. The purpose behind doing so would be to determine whether including the time walking on the treadmill actually improved model fit, after the penalty for model complexity was applied.

```
coronary.logistic.null<-glm(group~1, family=binomial)
```

The AIC for this intercept-only model was 29.726, which is larger than the 16.966 for the model including time. Based on AIC, along with the hypothesis test results discussed above, we would therefore conclude that the full model including time provided better fit to the outcome of coronary artery disease.

Logistic Regression Model for an Ordinal Outcome Variable

In the prior example, we considered an outcome variable that could take two possible values: 0 (healthy heart), 1 (diseased heart). In many instances a categorical outcome variable will have more than two potential outcomes, however. In this section we demonstrate the case where the dependent variable is ordinal in nature so that the categories can be interpreted as going from less to more, or smaller to larger (or vice versa). In the next section we will work with models that allow the categories to be unordered in nature.

As a way to motivate our discussion of ordinal logistic regression models, let's consider the following example. A dietician has developed a behavior-management system for individuals suffering from obesity that is designed to encourage a healthier lifestyle. One such healthy behavior is the preparation of their own food at home using fresh ingredients rather than dining out or eating prepackaged foods. Study participants consisted of 100 individuals who were under a physician's care for a health issue directly related to obesity. Members of the sample were randomly assigned to either a control condition in which they received no special instruction in how to plan and prepare healthy meals from scratch, or a treatment condition in which they did receive such instruction. The outcome of interest was a rating provided two months after the study began in which each subject indicated the extent to which they prepared their own meals. The response scale ranged from 0 (Prepared all of my own meals from scratch) to 4 (Never prepared any of my own meals from scratch), so that lower values were indicative of a stronger predilection to prepare meals at home from scratch. The dietician is interested in whether there are differences in this response between the control and treatment groups.

One commonly used model for ordinal data such as these is the cumulative logits model, which is as expressed as:

$$\text{logit}[P(Y \le j)] = \ln\left(\frac{P(Y \le j)}{1 - P(Y \le j)}\right). \tag{7.2}$$

In this model, there are J-1 logits where J is the number of categories in the dependent variable, and Y is the actual outcome value. Essentially, this model compares the likelihood of the outcome variable taking a value of j or lower, versus outcomes larger than j. For the current example there would be four separate logits:

$$\ln\left(\frac{p(Y = 0)}{p(Y = 1) + p(Y = 2) + p(Y = 3) + p(Y = 4)}\right) = \beta_{01} + \beta_1 x$$

$$\ln\left(\frac{p(Y = 0) + p(Y = 1)}{p(Y = 2) + p(Y = 3) + p(Y = 4)}\right) = \beta_{02} + \beta_1 x$$

$$\ln\left(\frac{p(Y = 0) + p(Y = 1) + p(Y = 2)}{p(Y = 3) + p(Y = 4)}\right) = \beta_{03} + \beta_1 x \tag{7.3}$$

$$\ln\left(\frac{p(Y = 0) + p(Y = 1) + p(Y = 2) + p(Y = 3)}{p(Y = 4)}\right) = \beta_{04} + \beta_1 x$$

In the cumulative logits model there is a single slope relating the independent variable to the ordinal response, and each logit has a unique intercept. In order for a single slope to apply across all logits we must make the proportional odds assumption, which states that this slope is identical across logits. In order to fit the cumulative logits model to our data in R, we use the polr function, as in this example.

```
cooking.cum.logit<-polr(cook~treatment, method=c("logistic"))
```

The dependent variable, cook, must be an R factor object, and the independent variable can be either a factor or numeric. In this case, treatment is coded as 0 (control) and 1 (treatment). To ensure that cook is a factor, we use cook<-as.factor(cook) prior to fitting the model. Using summary(cooking.cum.logit) after fitting the model, we obtain the following output.

```
Call:
polr(formula = cook ~ treatment, method = c("logistic"))
Coefficients:
                Value       Std. Error   t value
treatment -0.7963096   0.3677003  -2.165649
Intercepts:
    Value    Std. Error  t value
0|1 -2.9259  0.4381     -6.6783
1|2 -1.7214  0.3276     -5.2541
2|3 -0.2426  0.2752     -0.8816
3|4 1.3728   0.3228      4.2525
Residual Deviance: 293.1349
AIC: 303.1349
```

After the function call, we see the results for the independent variable, treatment. The coefficient value is −0.796, indicating that a higher value on the treatment variable (i.e. treatment=1) was associated with a greater likelihood of providing a lower response on the cooking item. Remember that lower responses to the cooking item reflected a greater propensity to eat scratch-made food at home. Thus, in this example those in the treatment conditions had a greater likelihood of eating scratch-made food at home. Adjacent to the coefficient value is the standard error for the slope, which is divided into the coefficient in order to obtain the *t*-statistic residing in the final column. We note that there is not a *p*-value associated with this *t*-statistic, because in the generalized linear model context this value only follows the *t* distribution asymptotically (i.e. for large samples). In other cases, it is simply an indication of the relative magnitude of the relationship between the treatment and outcome variable. In this context, we might consider the relationship to be "significant" if the *t*-value exceeds 2, which is approximately the *t* critical value for a two-tailed hypothesis test with $\alpha=0.05$ and infinite degrees of freedom. Using this criterion, we would conclude that there is indeed a statistically significant negative relationship between treatment condition and self-reported cooking behavior. Furthermore, by exponentiating the slope we can also calculate the relative odds of a higher-level response to the cooking item between the two groups. Much as we did in the dichotomous logistic regression case, we use the equation $e^{\beta 1}$ to convert the slope to an odds ratio. In this case, the value is 0.451, indicating that the odds of a treatment group member selecting a higher-level response (less self-cooking behavior) is only 0.451 as large as that of the control group. Note that this odds ratio applies to any pair of adjacent categories, such as 0 versus 1, 1 versus 2, 2 versus 3, or 3 versus 4.

R also provides the individual intercepts along with the residual deviance, and AIC for the model. The intercepts are, as with dichotomous logistic regression, the log odds of the target response when the independent variable is 0. In this example, a treatment of 0 corresponds to the control group. Thus, the intercept represents the log odds of the target response for the control condition. As we saw above, it is possible to convert this to the odds scale through exponentiating the estimate. The first intercept provides the log odds of a response of 0 versus all other values for the control group; i.e. plans and prepares all meals for oneself versus all other options. The intercept for this logit is −2.9259, which yields an $e^{-2.9259}$ of 0.054. We can interpret this to mean that the odds of a member of the control group planning and preparing their own meals versus something less are 0.054. In other words, it's highly unlikely a member of the control group will do this.

We can use the deviance, along with the appropriate degrees of freedom, to obtain a test of the null hypothesis that the model fits the data. The following command line in R will do this for us.

```
1-pchisq(deviance(cooking.cum.logit), df.residual(cooking.cum.
logit))
```

[1] 0

The p-value is extremely small (rounded to 0), indicating that the model as a whole does not provide very good fit to the data. This could mean that to obtain better fit, we need to include more independent variables with a strong relationship to the dependent. However, if our primary interest is in determining whether there are treatment differences in cooking behavior, then this overall test of model fit may not be crucial, since we are able to answer the question regarding the relationship of treatment to the cooking behavior item.

Multinomial Logistic Regression

A third type of categorical outcome variable is one in which there are more than two categories but the categories are not ordered. An example can be seen in a survey of likely voters who were asked to classify themselves as liberal, moderate, or conservative. A political scientist might be interested in predicting an individual's political viewpoint as a function of his/her age. The most common statistical approach for doing so is the generalized logits, or multinomial logistic regression model. This approach, which Agresti (2002) refers to as the baseline category logit model, assigns one of the dependent variable categories to be the baseline against which all other categories are compared. More formally, the multinomial logistic regression model can be expressed as

$$\ln\left(\frac{p(Y=i)}{p(Y=j)}\right) = \beta_{i0} + \beta_{i1}x \tag{7.4}$$

In this model, category j will always serve as the reference group against which the other categories, i, are compared. There will be a different logit for each non-reference category, and each of the logits will have a unique intercept (β_{i0}) and slope (β_{i1}). Thus, unlike with the cumulative logits model in which a single slope represented the relationship between the independent variable and the outcome, in the multinomial logits model we have multiple slopes for each independent variable, one for each logit. Therefore, we do not need to make the proportional odds assumption, which makes this model a useful alternative to the cumulative logits model when that assumption is not tenable. The disadvantage of using the multinomial logits model with an ordinal outcome variable is that the ordinal nature of the data is ignored. Any of the categories can serve as the reference, with the decision being based on the research question of most interest (i.e. against which group would comparisons be most interesting), or on pragmatic concerns such as which group is the largest, should the research questions not serve as the primary deciding factor. Finally, it is possible to compare the results for two non-reference categories using the equation

$$\ln\left(\frac{p(Y=i)}{p(Y=m)}\right) = \ln\left(\frac{p(Y=i)}{p(Y=j)}\right) - \ln\left(\frac{p(Y=m)}{p(Y=j)}\right). \tag{7.5}$$

For the present example, we will set the conservative group to be the reference, and fit a model in which age is the independent variable and political viewpoint is the dependent. In order to do so, we will use the function mulitnom within the nnet package, which will need to be installed prior to running the analysis. We would then use the command library(nnet) to make the functions in this library available. The data were read into the R data frame politics, containing the variables age and viewpoint, which were coded as C (conservative), M (moderate), or L (liberal) for each individual in the sample. Age was expressed as the number of years of age. The R command to fit the multinomial logistic regression model is politics. multinom<-multinom(viewpoint~age, data=politics), producing the following output.

```
# weights: 9 (4 variable)
initial value 1647.918433
final value 1617.105227
converged
```

This message simply indicates the initial and final values of the maximum likelihood fitting function, along with the information that the model converged. In order to obtain the parameter estimates and standard errors, we use the summary(politics.multinom).

```
Call:
multinom(formula = viewpoint ~ age, data = politics)
Coefficients:
          (Intercept)              age
L          0.4399943  -0.016611846
M          0.3295633  -0.004915465
Std. Errors:
          (Intercept)              age
L          0.1914777   0.003974495
M          0.1724674   0.003415578
Residual Deviance: 3234.210
AIC: 3242.210
```

Based on these results, we see that the slope relating age to the logit comparing self-identification as liberal (L) is –0.0166, indicating that older individuals had a lower likelihood of being liberal versus conservative. In order to determine whether this relationship is statistically significant, we can calculate a 95% confidence interval using the coefficient and the standard error for this term. This interval is constructed as

$$-0.0166 \pm 2(0.0040)$$
$$-0.0166 \pm 0.008$$
$$(-0.0246, -0.0086).$$

Because 0 is not in this interval, it is not a likely value of the coefficient in the population, leading us to conclude that the coefficient is statistically significant. In other words, we can conclude that in the population, older individuals are less likely to self-identify as liberal than as conservative. We can also construct a confidence interval for the coefficient relating age to the logit for moderate to conservative:

$$-0.0049 \pm 2(0.0034)$$
$$-0.0049 \pm 0.0068$$
$$(-0.0117, 0.0019).$$

Thus, because 0 does lie within this interval, we cannot conclude that there is a significant relationship between age and the logit. In other words, age is not related to the political viewpoint of an individual when it comes to comparing moderate versus conservative. Finally, we can calculate estimates for comparing L and M by applying Equation (7.5).

$$\ln\left(\frac{p(Y=L)}{p(Y=M)}\right) = \ln\left(\frac{p(Y=L)}{p(Y=C)}\right) - \ln\left(\frac{p(Y=M)}{p(Y=C)}\right)$$
$$= (0.4400 - 0.0166\,(\text{age})) - (0.3300 - 0.0049\,(\text{age}))$$
$$= 0.4400 - 0.3300 - 0.0166\,(\text{age}) + 0.0049\,(\text{age})$$
$$= 0.1100 - 0.0117\,(\text{age})$$

When these are taken together, we would conclude that older individuals are less likely to be liberal than conservative, and less likely to be liberal than moderate.

Models for Count Data

Poisson Regression

To this point, we have been focused on outcome variables of a categorical nature, such as whether an individual cooks for themselves, or the presence/absence of coronary artery disease. Another type of data that does not fit nicely into the standard models assuming normally distributed errors involves counts or rates of some outcome, particularly of rare events. Such variables often follow the Poisson distribution, a major property of which is that the mean is equal to the variance. It is clear that if the outcome variable is a count, its lower bound must be 0; i.e. one cannot have negative counts. This presents a problem to researchers applying the standard linear regression model, as it may produce predicted values of the outcome that are less than 0, and thus are nonsensical. In order to deal with this potential difficulty, Poisson regression was developed. This approach to dealing with count data rests upon the application of the log to the outcome variable, thereby overcoming the problem of negative predicted counts, since the log of the outcome can take any real number value. Thus, when dealing with the Poisson distribution in the form of counts, we will use the log as the link function in fitting the Poisson regression model:

$$\ln(Y) = \beta_0 + \beta_1 x. \tag{7.6}$$

In all other respects, the Poisson model is similar to other regression models in that the relationship between the independent and dependent variables is expressed through the slope, β_1. And again, the assumption underlying the Poisson model is that the mean is equal to the variance. This assumption is typically expressed by stating that the overdispersion parameter $\phi = 1$. The ϕ parameter appears in the Poisson distribution density and thus is a key component in the fitting function used to determine the optimal model parameter estimates in maximum likelihood. A thorough review of this fitting function is beyond the scope of this book. Interested readers are referred to Agresti (2002) for a complete presentation of this issue.

Estimating the Poisson regression model in R can be done with the GLM function that we used previously for dichotomous logistic regression. Consider an example in which a demographer is interested in determining whether there exists a relationship between the socioeconomic status (sei) of a family and the number of children under the age of six months (babies) who are living in the home. We first read the data and name it ses _ babies. We then attach it using attach(ses _ babies). In order to see the distribution of the number of babies, we can use the command hist(babies), with the resulting histogram appearing below.

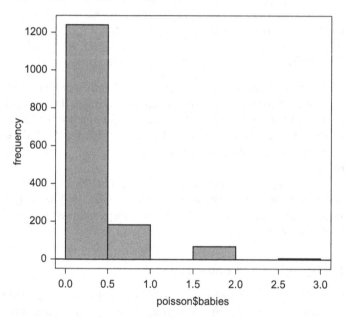

We can see that 0 was the most common response by individuals in the sample, with the maximum number being 3.

In order to fit the model with the glm function, we would use the following function call.

```
babies.poisson<-glm(babies~sei, data= ses_babies,
family=c("poisson"))
```

In this command sequence, we create an object called babies.poisson, which includes the output for the Poisson regression model. The function call is identical to that used for nearly all models in R, and we define the distribution of the outcome variable in the family statement. Using the summary(babies.poisson) yields the following output.

```
Call:
glm(formula = babies ~ sei, family = c("poisson"), data =
ses_babies)
Deviance Residuals:
    Min       1Q    Median       3Q       Max
-0.7312  -0.6914  -0.6676  -0.6217    3.1345
Coefficients:
            Estimate Std. Error z value Pr(>|z|)
(Intercept) -1.268353   0.132641  -9.562   <2e-16 ***
sei         -0.005086   0.002900  -1.754   0.0794 .
---
Signif. codes: 0 '***' 0.001 '**' 0.01 '*' 0.05 '.' 0.1 ' ' 1
(Dispersion parameter for poisson family taken to be 1)
 Null deviance: 1237.8 on 1496 degrees of freedom
Residual deviance: 1234.7 on 1495 degrees of freedom
 (3 observations deleted due to missingness)
AIC: 1803
Number of Fisher Scoring iterations: 6
```

These results show that sei did not have a statistically significant relationship with the number of children under six months old living in the home ($p = 0.0794$). We can use the following command to obtain the p-value for the test of the null hypothesis that the model fits the data. 1-pchisq(deviance(babies.poisson), df.residual(babies.poisson))

[1] 0.9999998

The resulting p is clearly not significant at $\alpha = 0.05$, suggesting that the model does appear to fit the data adequately. The AIC of 1,803 will be useful as we compare the relative fit of the Poisson regression model with that of other models for count data.

Models for Overdispersed Count Data

Recall that a primary assumption underlying the Poisson regression model is that the mean and variance are equal. When this assumption does not hold, such as when the variance is larger than the mean, estimation of model standard errors is compromised so that they tend to be smaller than is actually true in the population (Agresti, 2002). For this reason, it is important that researchers dealing with count data investigate whether this key assumption is likely to hold in the population. Perhaps the most direct way to do this is to fit alternative models that relax the $\phi = 1$ restriction that we had in Poisson regression. One approach for doing so is to use the quasipoisson model. The quasipoisson takes the same form as the Poisson regression model, but does not constrain ϕ to be 1, which in turn will lead to different standard errors for the parameter estimates, though the values of the coefficient estimates themselves will not change. The quasipoisson model can be fit in R using the glm function, with the family set to quasipoisson: babies.

```
quasipoisson<-glm(babies~sei, data= ses_babies, family=
c("quasipoisson")).
```
We can obtain the output using the summary function.

```
Call:
glm(formula = babies ~ sei, family = c("quasipoisson"), data =
ses_babies)

Deviance Residuals:
    Min      1Q   Median      3Q      Max
-0.7312  -0.6914  -0.6676  -0.6217   3.1345
Coefficients:
            Estimate Std. Error t value Pr(>|t|)
(Intercept) -1.268353   0.150108   -8.45  <2e-16 ***
sei         -0.005086   0.003282   -1.55   0.121
---
Signif. codes: 0 '***' 0.001 '**' 0.01 '*' 0.05 '.' 0.1 ' ' 1
(Dispersion parameter for quasipoisson family taken to be
1.280709)
    Null deviance: 1237.8 on 1496 degrees of freedom
Residual deviance: 1234.7 on 1495 degrees of freedom
 (3 observations deleted due to missingness)
AIC: NA
Number of Fisher Scoring iterations: 6
```

As noted above, the coefficients themselves are the same in the quasipoisson and Poisson regression models. However, the standard errors in the former are somewhat larger than those in the latter. In addition, the estimate of ϕ is provided for the quasipoisson model, and is 1.28 in this case. While this is not exactly equal to 1, it is also not markedly larger, suggesting that the data are not terribly overdispersed. We can test for model fit as we did with the Poisson regression, using the command

```
1-pchisq(deviance(babies.quasipoisson), df.residual(babies.
quasipoisson))
```

[1] 0.9999998

And, as with the Poisson, the quasipoisson model also fits the data adequately.

A second alternative to the Poisson model when data are overdispersed is a regression model based on the negative binomial distribution. The mean of the negative binomial distribution is identical to that of the Poisson, while the variance is

$$\text{var}(Y) = \mu + \frac{\mu^2}{\theta}. \tag{7.7}$$

From Equation (7.7), it is clear that as θ increases in size, the variance approaches the mean, and the distribution becomes more like the Poisson. It is possible for the researcher to provide a value for θ if it is known that the data come from a particular distribution with a known θ. For example, when $\theta = 1$, the data are modeled from the gamma distribution. However, for most applications, the distribution is not known, in which case θ will be estimated from the data.

The negative binomial distribution can be fit to the data in R using the glm.nb function that is part of the MASS library. For the current example, the R commands to fit the negative binomial model and obtain the output would be

```
babies.nb<-glm.nb(babies~sei, data= ses_babies)
summary(babies.nb)


Call:
glm.nb(formula = babies ~ sei, data = ses_babies, init.theta =
0.60483559440229,
    link = log)
Deviance Residuals:
    Min       1Q  Median       3Q      Max
-0.6670  -0.6352  -0.6158  -0.5778   2.1973

Coefficients:
            Estimate Std. Error z value Pr(>|z|)
(Intercept) -1.260872   0.156371  -8.063  7.42e-16 ***
sei         -0.005262   0.003386  -1.554     0.120
---
Signif. codes: 0 '***' 0.001 '**' 0.01 '*' 0.05 '.' 0.1 ' ' 1

(Dispersion parameter for Negative Binomial(0.6048) family
taken to be 1)
    Null deviance: 854.08 on 1496 degrees of freedom
Residual deviance: 851.72 on 1495 degrees of freedom
 (3 observations deleted due to missingness)
AIC: 1755.4
Number of Fisher Scoring iterations: 1
        Theta: 0.605
    Std. Err.: 0.127

 2 x log-likelihood: -1749.395
```

Just as we saw with the quasipoisson regression, the parameter estimates for the negative binomial regression are identical to those for the Poisson. This fact simply reflects the common mean that the distributions all share. However, the standard errors for the estimates differ across the three models, although those for the negative binomial are very similar to those from the quasipoisson. Indeed, the resulting hypothesis test results provide the same answer for all three models, that there is not a statistically significant relationship between the sei and the number of babies living in the home. In addition to the parameter estimates and standard errors, we also obtain an estimate of θ of 0.605. In terms of determining which model is optimal, we can compare the AIC from the negative binomial (1755.4) to that of the Poisson (1803), to conclude that the former provides somewhat better fit to the data than the latter. In short, it would appear that the data are somewhat overdispersed as the model designed to account for this (negative binomial) provides better fit than the Poisson, which assumes no overdispersion. From a more practical perspective, the results of the two models are very similar, and a researcher using α = 0.05 would reach the same conclusion regarding the lack of relationship between sei and the number of babies living in the home, regardless of which model they selected.

Summary

Chapter 7 marks a major change in direction in terms of the type of data upon which we will focus. Through the first six chapters, we have been concerned with models in which the dependent variable is continuous, and generally assumed to be normally distributed. In Chapter 7 we learned about a variety of models designed for categorical dependent variables. In perhaps the simplest instance, such variables can be dichotomous, so that logistic regression is most appropriate for data analysis. When the outcome variable has more than two ordered categories, we see that logistic regression can be easily extended in the form of the cumulative logits model. For dependent variables with unordered categories, the multinomial logits model is the typical choice, and can be easily employed with R. Finally, we examined dependent variables that are counts, in which case we may choose the Poisson regression, the quasipoisson model, or the negative binomial model, depending upon how frequently the outcome being counted occurs. As with Chapter 1, the goal of Chapter 7 was primarily to provide an introduction to the single-level versions of the multilevel models to come. In Chapter 8 we will see that the model types described here can be extended into the multilevel context using our old friends lme and lmer.

8

Multilevel Generalized Linear Models (MGLMs)

In the previous chapter, we introduced generalized linear models, which are useful when the outcome variable of interest is categorical in nature. We described a number of models in this broad family, including logistic regression for binary, ordinal, and multinomial data distributions, as well as Poisson regression models for count, or frequency data. In each of those examples, the data were collected at a single level. However, just as is true for normally distributed outcome variables, it is common for categorical variables to be gathered in a multilevel framework. The focus of this chapter is on models designed specifically for scenarios in which the outcome of interest is either categorical or count in nature, and the data have been collected in a multilevel framework. Chapter organization will mirror that of Chapter 7, beginning with a description of fitting logistic regression for dichotomous data, followed by models for ordinal and nominal dependent variables, and concluding with models for frequency count data that fit the Poisson distribution, and for the case of overdispersed counts. Given that the previous chapter provided the relevant mathematical underpinnings for these various models in the single-level case, and Chapter 2 introduced some of the theory underlying multilevel models, the current chapter will focus almost exclusively on the application of the R software package to fit these models, and on the interpretation of the resultant output.

MGLMs for a Dichotomous Outcome Variable

In order to introduce MGLMs for dichotomous outcomes, let us consider the following example. A researcher has collected testing data indicating whether 9,316 students have passed a state mathematics assessment, along with several measures of mathematics aptitude that were measured prior to the administration of the achievement test. She is interested in whether there exists a relationship between the score on number sense aptitude and the likelihood that a student will achieve a passing score on the mathematics achievement test, for which all examinees are categorized as either passing (1) or failing (0). Given that the outcome variable is dichotomous, we could use the binary logistic regression method introduced in Chapter 7. However, students in this sample are clustered by school, as was the case with the data that were examined in Chapters 3 and 4. Therefore, we will need to appropriately account for this multilevel data structure in our regression analysis.

Random Intercept Logistic Regression

In order to fit the random intercept logistic regression model in R, we will use the glmer function for fitting GLiMs that are associated with the lme4 library. The algorithm underlying glmer relies on an adaptive Gauss-Hermite likelihood approximation (Liu and Pierce, 1994) to fit the model to the data. Given the applied focus of this book, we will not devote time to the technical specifications of this method for fitting the model to the data. However, the interested reader is referred to the Liu and Pierce work for a description of this method. The R command for fitting the model and obtaining the summary statistics appears below, following the call to the lme4 library and the attaching of the file containing the data. Note that in this initial analysis, we have a fixed effect for the intercept and the slope of the independent variable numsense, but we only allow a random intercept, thereby assuming that the relationship between the number sense score and the likelihood of achieving a passing score on the state math assessment (score2) is fixed across schools; i.e. the relationship of numsense with score2 does not vary from one school to another.

```
library(lme4)
attach(mathfinal)
summary(model8.1<-glmer(score2~numsense+(1|school),family=bino
mial, na.action=na.omit))

Generalized linear mixed model fit by maximum likelihood
(Laplace Approximation) ['glmerMod']
 Family: binomial ( logit )
Formula: score2 ~ numsense + (1 | school)

    AIC     BIC   logLik  deviance df.resid
 9835.9  9857.4  -4915.0    9829.9      9313

Scaled residuals:
    Min      1Q  Median      3Q     Max
-5.2722 -0.7084  0.2870  0.6448  3.4279

Random effects:
 Groups Name         Variance Std.Dev.
 school (Intercept)  0.2888   0.5374
Number of obs: 9316, groups:  school, 40
```

```
Fixed effects:
        Estimate Std. Error z value Pr(>|z|)
(Intercept) -11.659653  0.305358 -38.18  <2e-16 ***
numsense     0.059177  0.001446  40.94  <2e-16 ***
---
Signif. codes: 0 '***' 0.001 '**' 0.01 '*' 0.05 '.' 0.1 ' ' 1

Correlation of Fixed Effects:
         (Intr)
numsense -0.955
```

The function call is similar to what we saw with linear models in Chapter 3. In terms of interpretation of the results, we first examine the variability in intercepts from school to school. This variation is presented as both the variance and standard deviation of the U_{0j} terms from Chapter 2 (τ_0^2), which are 0.2888 and 0.5374, respectively, for this example. The modal value of the intercept across schools is −11.659653. With regard to the fixed effect, the slope of numsense, we see that higher scores are associated with a greater likelihood of passing the state math assessment, with the slope being 0.059177 ($p < .05$). (Remember that R models the larger value of the outcome in the numerator of the logit, and in this case, passing was coded as 1 whereas failing was coded as 0.) The standard error, test statistic, and p-value appear in the next three columns. The results are statistically significant ($p < 0.001$), leading to the conclusion that overall, number sense scores are positively related to the likelihood of a student achieving a passing score on the assessment. Finally, we see that the correlation between the slope and intercept is strongly negative (−0.955). Given that this is an estimate of the relationship between two fixed effects, we are not particularly interested in it. Information about the residuals appears at the very end of the output.

As we discussed in Chapter 3, it is useful for us to obtain confidence intervals for parameter estimates of both the fixed and random effects in the model. With lme4, we have several options in this regard, using the confint function, as we saw in Chapter 3. Indeed, precisely the same options are available for multilevel generalized models fit using glmer, as was the case for models fit using the lmer function. We would refer the reader to Chapter 3 for a review of how each of these methods works. In the following text, we demonstrate the use of each of these confidence interval types for model8.1, and then discuss implications of these results for interpretation of the problem at hand.

```
#Percentile Bootstrap

confint(model8.1, method=c("boot"), boot.type=c("perc"))
Computing bootstrap confidence intervals ...
                 2.5 %        97.5 %
.sig01         0.40901802    0.66941398
(Intercept)  -12.25445259  -11.08224166
numsense       0.05649295    0.06206932
```

```
#Basic Bootstrap

confint(model8.1, method=c("boot"), boot.type=c("basic"))
Computing bootstrap confidence intervals ...
                       2.5 %        97.5 %
.sig01           0.41170646    0.69754346
(Intercept)    -12.23664456  -11.03448543
numsense         0.05603513    0.06197624

#Normal Bootstrap

confint(model8.1, method=c("boot"), boot.type=c("norm"))
Computing bootstrap confidence intervals ...
                       2.5 %        97.5 %
.sig01           0.41896917     0.6886556
(Intercept)    -12.25037745   -11.0649487
numsense         0.05641875     0.0619302

#Wald

confint(model8.1, method=c("Wald"))
                       2.5 %        97.5 %
.sig01                    NA            NA
(Intercept)    -12.25814446  -11.06116177
numsense         0.05634366    0.06201022

#Profile

confint(model8.1, method=c("profile"))
Computing profile confidence intervals ...
                       2.5 %        97.5 %
.sig01           0.42422171     0.6968439
(Intercept)    -12.26384797   -11.0669380
numsense         0.05637074     0.0620358
```

For the random effect, all of the methods for calculating confidence intervals yield similar results, in that the lower bound is approximately 0.41 to 0.42 and the upper bound is between 0.67 and 0.70. Regardless of the method, we see that 0 is not in the interval, and thus would conclude that the random intercept variance is statistically significant; i.e. there are differences in the intercept across schools. Similarly, the confidence intervals for the fixed effects (intercept and the coefficient for numsense) also did not include 0, which indicates that they were statistically significant as well. In particular, this would lead us to the conclusion that there is a positive relationship between numsense and the likelihood of receiving a passing test score, which we noted above.

In addition to the model parameter estimates, the results from glmer also include information about model fit, in particular values for the AIC and

BIC. As we have discussed in previous chapters, these statistics can be used to compare the relative fit of various models in an attempt to pick the optimal one, with smaller values indicating better model fit. As well as comparing relative model fit, the fit of two nested models created by glmer can also be compared with one another in the form of a likelihood ratio test with the anova function. The null hypothesis of this test is that the fit of two nested models is equivalent, so that a statistically significant result (i.e. $p \leq 0.05$) would indicate that the models provide different fit to the data, with the more complicated (fuller) model typically providing an improvement in fit beyond what would be expected with the additional parameters added. We will demonstrate the use of this test in the next section.

Random Coefficients Logistic Regression

As with the linear multilevel models, it is also possible to allow for random slopes with multilevel GLiMs, and this can be done using glmer, as below.

```
summary(model8.2<-glmer(score2~numsense+(numsense|school),fami
ly=binomial))

Generalized linear mixed model fit by maximum likelihood
(Laplace Approximation) ['glmerMod']
 Family: binomial ( logit )
Formula: score2 ~ numsense + (numsense | school)

    AIC      BIC   logLik deviance df.resid
 9768.7   9804.4  -4879.4   9758.7     9311

Scaled residuals:
    Min      1Q   Median      3Q      Max
-4.7472 -0.6942   0.2839  0.6352   3.6943

Random effects:
 Groups Name          Variance  Std.Dev. Corr
 school (Intercept)   2.170e+01 4.65843
        numsense      4.105e-04 0.02026  -1.00
Number of obs: 9316, groups:  school, 40

Fixed effects:
              Estimate Std. Error z value Pr(>|z|)
(Intercept) -12.901304   0.836336  -15.43   <2e-16 ***
numsense      0.064903   0.003735   17.38   <2e-16 ***
---
Signif. codes:  0 '***' 0.001 '**' 0.01 '*' 0.05 '.' 0.1 ' ' 1

Correlation of Fixed Effects:
         (Intr)
numsense -0.995
```

We will focus on aspects of the output for the random coefficients model that differ from those of the random intercepts. In particular, note that we have an estimate of τ_1^2, the variance of the U_{1j} estimates for specific schools. This value, 0.00041, is relatively small when compared to the variation of intercepts across schools (0.217), meaning that the relationship of number sense with the likelihood of an individual receiving a passing score on the math achievement test is relatively similar across the schools. The modal slope across schools is 0.064274, again indicating that individuals with higher number sense scores also have a higher likelihood of passing the math assessment. Finally, it is important to note that the correlation between the random components of the slope and intercept, the standardized version of τ_{01}, is very strongly negative.

As with the random intercept model, we can obtain confidence intervals for the random and fixed effects of the random coefficients model. The same options are available as was the case for the random intercept model. For this example, we will use the Wald, profile, and percentile bootstrap confidence intervals.

```
#Wald

confint(model8.2, method=c("Wald"))
                  2.5 %            97.5 %
.sig01               NA                NA
.sig02               NA                NA
.sig03               NA                NA
(Intercept)  -14.54049182  -11.26211656
numsense       0.05758284    0.07222221


#Profile
confint(model8.2, method=c("profile"))
Computing profile confidence intervals ...
                  2.5 %            97.5 %
.sig01        3.33785318    6.47047746
.sig02       -0.99922647   -0.98878396
.sig03        0.01428772    0.02843558
(Intercept) -14.66242289  -11.28916466
numsense      0.05768995    0.07285651


#Percentile Bootstrap
confint(model8.2, method=c("boot"), boot.type=c("perc"))
Computing bootstrap confidence intervals ...
                  2.5 %            97.5 %
.sig01        3.18387655    6.01887096
.sig02       -0.99869150   -0.99035490
.sig03        0.01367105    0.02621692
(Intercept) -14.88458123  -11.29933632
numsense      0.05781441    0.07362581
```

None of the confidence intervals for the random effects included 0, leading us to conclude that each of them is likely to be different from 0 in the population.

The inclusion of AIC and BIC in the GLMER output allows for a direct comparison of model fit, thus aiding in the selection of the optimal model for the data. As a brief reminder, AIC and BIC are both measures of unexplained variation in the data with a penalty for model complexity. Therefore, models with lower values provide relatively better fit. Comparison of either AIC or BIC between Models 8.1 (AIC = 9835.9, BIC = 9857.4) and 8.2 (AIC = 9768.7, BIC = 9804.4) reveals that the latter provides better fit to the data. We do need to remember that AIC and BIC are not significance tests, but rather measures of relative model fit. In addition to the relative fit indices, we can also compare the fit of the two models using the anova command, as we demonstrated in Chapter 3.

```
anova(model8.1, model8.2)

Data: NULL
Models:
model8.1: score2 ~ numsense + (1 | school)
model8.2: score2 ~ numsense + (numsense | school)
        Df    AIC    BIC  logLik deviance  Chisq Chi Df
Pr(>Chisq)
model8.1  3 9835.9 9857.4 -4915.0   9829.9
model8.2  5 9768.7 9804.4 -4879.4   9758.7 71.208      2
3.447e-16 ***
---
Signif. codes:  0 '***' 0.001 '**' 0.01 '*' 0.05 '.' 0.1 ' ' 1
```

These results indicate that there is a statistically significant difference in the relative fit of the two models. Furthermore, the AIC and BIC are both lower for model8.2, suggesting that it provides better fit to the data than does model8.2. Thus, we can conclude that the coefficients for numsense are significantly different across schools, so that allowing them to vary among the schools leads to a more optimal model than forcing them to be the same. It should be noted that this comparison of model fit carried out by the anova command relies on a 2-degrees of freedom test, in which a significant difference in fit may be due to the fixed effects, the random effects, or a combination of the two. Another way to interpret this result is that there do appear to be school differences in the relationship of number sense and the likelihood of students passing the mathematics achievement test.

Inclusion of Additional Level-1 and Level-2 Effects in MGLM

The researcher in our example is also interested in learning whether there is a statistically significant relationship between an examinee's gender (female, where 1 = female, and 0 = male) and the likelihood of passing the state math assessment, as well as the relationship of passing and number sense score. In addition, we include a level-2 predictor, the proportion of students in the school receiving free lunch. In order to fit the additional level-1 variable to the random coefficients model, we would use the following command to obtain the subsequent output. Note that this is a random intercept model, with no random coefficients.

```
summary(model8.3<-glmer(score2~numsense+female+L_
Free+(1|school), family=binomial), data=mathfinal.nomiss,
na.action=na.omit)

Generalized linear mixed model fit by maximum likelihood
(Laplace Approximation) ['glmerMod']
 Family: binomial  ( logit )
Formula: score2 ~ numsense + female + L_Free + (1 | school)

     AIC       BIC   logLik deviance df.resid
  7064.0    7098.1  -3527.0   7054.0     6805

Scaled residuals:
    Min      1Q  Median      3Q     Max
-4.8524 -0.6875  0.2734  0.6301  3.7430

Random effects:
 Groups Name            Variance Std.Dev.
 school (Intercept) 0.2824     0.5314
Number of obs: 6810, groups:  school, 32

Fixed effects:
                 Estimate Std. Error z value Pr(>|z|)
(Intercept)    -11.969032   0.424692 -28.183   <2e-16 ***
numsense         0.063229   0.001761  35.911   <2e-16 ***
female          -0.021276   0.058786  -0.362   0.7174
L_Free          -0.008820   0.003733  -2.363   0.0181 *
---
Signif. codes:  0 '***' 0.001 '**' 0.01 '*' 0.05 '.' 0.1 ' ' 1

Correlation of Fixed Effects:
            (Intr) numsns mthf.$
numsense    -0.842
female      -0.068 -0.001
L_Free      -0.491  0.013 -0.005
```

These results indicate that being female is not significantly related to one's likelihood of passing the math achievement test; i.e. there are no gender differences in the likelihood of passing. Scores on the numsense subscale were positively associated with the likelihood of passing the test, and attending a school with a higher proportion of students on free lunch was associated with a lower such likelihood. The 95% bootstrap percentile confidence intervals for the fixed and random effects appear below. The interval for the random intercept effect does not include 0, meaning that this term is not likely to be 0 in the population. In other words, there are school to school differences in the likelihood of individuals receiving a passing test grade. In addition, intervals for the intercept, numsense score, and proportion of students on free lunch all excluded 0, reinforcing the hypothesis test results in the initial glmer output.

```
confint(model8.3, method=c("boot"), boot.type=c("perc"))

Computing bootstrap confidence intervals ...
                             2.5 %         97.5 %
.sig01                  0.35843219    0.67053482
(Intercept)           -12.87837065  -11.15704771
numsense                0.05991524    0.06721941
female                 -0.13122349    0.09747311
L_Free                 -0.01607477   -0.00119302
```

As was the case with model8.2, we can estimate a random coefficients model. In this case, we will estimate the random coefficient for female.

```
summary(model8.4<-glmer(score2~numsense+female+L_Free+(female|
school),family=binomial), data=mathfinal.nomiss, na.action=na.
omit)

Generalized linear mixed model fit by maximum likelihood
(Laplace Approximation) ['glmerMod']
 Family: binomial  ( logit )
Formula: score2 ~ numsense + female + L_Free + (mathfinal.
nomiss$female |
    school)

     AIC      BIC   logLik deviance df.resid
  7065.3   7113.1  -3525.7   7051.3     6803

Scaled residuals:
    Min      1Q  Median      3Q     Max
-4.6373 -0.6880  0.2752  0.6279  3.6198

Random effects:
 Groups Name                 Variance Std.Dev. Corr
 school (Intercept)          0.22883  0.4784
 female                      0.01029  0.1015   1.00
Number of obs: 6810, groups:  school, 32

Fixed effects:
```

	Estimate	Std. Error	z value	Pr(>|z|)
(Intercept)	-12.002707	0.422308	-28.422	<2e-16
***				
numsense	0.063286	0.001761	35.945	<2e-16
***				
female	-0.015689	0.061785	-0.254	0.7996
L_Free	-0.008498	0.003627	-2.343	0.0191
*				

```
---
Signif. codes:  0 '***' 0.001 '**' 0.01 '*' 0.05 '.' 0.1 ' ' 1

Correlation of Fixed Effects:
          (Intr) numsns mthf.$
numsense  -0.850
female    -0.043  0.004
L_Free    -0.492  0.021  0.065
```

The variance estimate for the random coefficient effect for gender is 0.01029. The coefficients for the other variables in model8.4 are quite similar to those in model8.3. In order to ascertain whether the random coefficient for `female` is statistically significant, we can examine the 95% confidence interval, in this case again using the percentile bootstrap.

```
confint(model8.4, method=c("boot"), boot.type=c("perc"))
```

```
Computing bootstrap confidence intervals ...
                               2.5 %          97.5 %
.sig01                     0.312070590    6.390120e-01
.sig02                    -0.999999715    1.000000e+00
.sig03                     0.008746154    2.609692e-01
(Intercept)              -12.840932662   -1.116645e+01
numsense                   0.059872387    6.694960e-02
mathfinal.nomiss$female   -0.160624776    1.160213e-01
L_Free                    -0.016086709   -7.227924e-04
```

These results show that the correlation between the fixed effects (.sig02) is not different from 0, nor is the coefficient linking `female` to the outcome variable, because in both cases the confidence interval includes 0.

We can use the relative fit indices to make some judgments regarding which of these two models might be optimal for better understanding the population. Both AIC and BIC are very slightly smaller for model8.3 (7064.0 and 7098.1), as compared to model8.4 (7065.3 and 7113.1), indicating that the simpler model (without the `female` random coefficient term) might be preferable. In addition, the results of the likelihood ratio test, which appear below, reveal that the fit of the two models is not statistically significantly different. Given this statistically equivalent fit, we would prefer the simpler model, without the random coefficient effect for `female`.

```
anova(model8.3, model8.4)
Data: NULL
Models:
model8.3: score2 ~ numsense + female + L_Free + (1 | school)
model8.4: score2 ~ numsense + female + L_Free + (female |
school)
        Df    AIC    BIC  logLik deviance  Chisq Chi Df
Pr(>Chisq)
model8.3  5 7064.0 7098.1 -3527.0   7054.0
model8.4  7 7065.3 7113.1 -3525.7   7051.3 2.6122      2
0.2709
```

MGLM for an Ordinal Outcome Variable

As was the case for non-multilevel data, the cumulative logits link function can be used with ordinal data in the context of multilevel logistic regression. Indeed, the link will be the familiar cumulative logit that we described in Chapter 7. Furthermore, the multilevel aspects of the model, including random intercept and coefficient, take the same form as what we described above. To provide context, let's again consider the math achievement results for students. In this case, the outcome variable takes one of three possible values for each member of the sample: 1=failure, 2=pass, 3=pass with distinction. In this case, the question of most interest to the researcher is whether a computation aptitude score is a good predictor of status on the math achievement test.

Random Intercept Logistic Regression

In order to fit a multilevel cumulative logits model using R, we install the ordinal package, which allows for fitting a variety of mixed effects models for categorical outcomes. Within this package, the clmm function is used to actually fit the multilevel cumulative logits model. Model parameter estimation is achieved using maximum likelihood based on the Newton-Raphson method. Once we have installed this package, we will use the library(ordinal) statement to load it. The R command to then fit the model and obtain the results, and the results themselves, appear below.

```
Summary(model8.5<-clmm(as.factor(score)~computation+(1|sch
ool), data=mathfinal, na.action=na.omit))

Cumulative Link Mixed Model fitted with the Laplace
approximation
```

```
formula: as.factor(score) ~ computation + (1 | school)
data:     mathfinal

 link  threshold nobs logLik    AIC       niter     max.grad
cond.H
 logit flexible  9316 -6678.87 13365.74 132(416)  7.10e-01
9.5e+05

Random effects:
 Groups Name           Variance Std.Dev.
 school (Intercept) 0.3844    0.62
Number of groups:   school 40

Coefficients:
            Estimate Std. Error z value Pr(>|z|)
computation  0.06977    0.00143   48.78   <2e-16 ***
Signif. Codes:  0 '***' 0.001 '**' 0.01 '*' 0.05 '.' 0.1 ' ' 1

Threshold coefficients:
    Estimate Std. Error z value
1|2  13.6531    0.3049   44.77
2|3  17.2826    0.3307   52.26
```

One initial point to note is that the syntax for clmm is very similar in form
to that for lmer. As with most R model syntax, the outcome variable (score)
is separated from the fixed effect (computation) by ~, and the random
effect, school, is included in parentheses along with 1, to denote that we are
fitting a random intercepts model. We should also note that the dependent
variable needs to be a factor, leading to our use of as.factor(score) in
the command sequence. It is important to state at this point that, currently,
there is not an R package available to fit a random coefficients model for the
cumulative logits model.

An examination of the results presented above reveals that the variance
and standard deviation of intercepts across schools are 0.3844 and 0.62,
respectively. Given that the variation is not near 0, we would conclude that
there appear to be differences in intercepts from one school to the next. In
addition, we see that there is a significant positive relationship between performance on the computation aptitude subtest and performance on the math
achievement test, indicating that examinees who have higher computation
skills also are more likely to attain higher ordinal scores on the achievement
test; e.g. pass versus fail or pass with distinction versus pass. We also obtain
estimates of the model intercepts, which are termed thresholds by clmm. As
was the case for the single-level cumulative logits model, the intercept represents the log odds of the likelihood of one response versus the other (e.g. 1
versus 2) when the value of the predictor variable is 0. A computation score
of 0 would indicate that the examinee did not correctly answer any of the
items on the test. Applying this fact to the first intercept presented above,

along with the exponentiation of the intercept that was demonstrated in the previous chapter, we can conclude that the odds of a person with a computation score of 1 passing the math achievement exam are $e^{13.6531} = 850,092.12$ to 1, or quite high! Finally, we also have available the AIC value (13365.74), which we can use to compare the relative fit of this to other models.

As an example of fitting models with both level-1 and level-2 variables, let's include the proportion of students receiving free lunch in the schools (L _ Free) as an independent variable along with the computation score.

```
summary(model8.6<-clmm(as.factor(score)~computation+L_
Free+(1|school), data=mathfinal, na.action=na.omit))

Cumulative Link Mixed Model fitted with the Laplace
approximation

formula: as.factor(score) ~ computation + L_Free + (1 |
school)
data:     mathfinal

 link   threshold nobs logLik    AIC       niter      max.grad
cond.H
 logit flexible   7069 -5035.13 10080.25 260(808) 1.54e-02
1.6e+06

Random effects:
 Groups Name           Variance Std.Dev.
 school (Intercept) 0.4028    0.6347
Number of groups:   school 34

Coefficients:
              Estimate Std. Error z value Pr(>|z|)
computation   0.074606  0.001698  43.940   <2e-16 ***
L_Free       -0.007612  0.004099  -1.857   0.0633 .
---
Signif. codes:  0 '***' 0.001 '**' 0.01 '*' 0.05 '.' 0.1 ' ' 1

Threshold coefficients:
    Estimate Std. Error z value
1|2  14.2381    0.4288   33.21
2|3  17.8775    0.4545   39.33
```

Given that we have already discussed the results of the previous model in some detail, we will not reiterate those basic ideas again. However, it is important to note those aspects that are different here. Specifically, the variability in the intercepts declined somewhat with the inclusion of the school-level variable, L _ Free. In addition, we also see that there is not a statistically significant relationship between the proportion of individuals receiving free lunch in the school and the likelihood that an individual student will obtain a higher achievement test score. Finally, a comparison of the AIC values for the computation-only model (13365.74) and the computation and free-lunch model (10080.25) shows that model8.6 provides a somewhat better fit to the data than does model8.5, given its smaller AIC value. In other words, in terms of the model fit to the data, we are better off including both free-lunch and computation score when modeling the three-level achievement outcome variable, even though L _ Free is statistically significantly related to the outcome variable. Note that the anova command is not available for models fit with clmm.

As of the writing of this book (December 2018), lme4 does not provide for the fitting of multilevel ordinal logistic regression models. Therefore, the clmm function within the ordinal package represents perhaps the most straightforward mechanism for fitting such models, albeit with its own limitations. As can be seen above, the basic fitting of these models is not complex, and indeed the syntax is similar to that of lme4. In addition, the ordinal package also allows for the fitting of ordered outcome variables in the non-multilevel context (see the clm function), and for multinomial outcome variables (see the clmm2 function, discussed below). As such, it represents another method available for fitting such models in a unified framework.

MGLM for Count Data

In the previous chapter, we examined statistical models designed for use with outcome variables that represented the frequency of some event occurring. Typically, these events were relatively rare, such as the number of babies in a family. Perhaps the most common distribution associated with such counts is the Poisson, a distribution in which the mean and the variance are equal. However, as we saw in Chapter 7, this equality of the two moments does not always hold in all empirical contexts, in which case we have what is commonly referred to as overdispersed data. In such cases, the Poisson regression model relating one or more independent variables to a count-dependent variable is not appropriate, and we must make use of either the quasipoissson or negative binomial distributions, each of which is able to appropriately model the inequality of the mean and variance. It is a fairly straightforward matter to extend any of these models to the multilevel context, both conceptually and using R with the appropriate packages. In the following sections, we will demonstrate analysis of multilevel count data outcomes in the context of Poisson regression, quasipoisson regression, and negative binomial regression in R. The example to be used involves the number of cardiac warning incidents (e.g. chest pain, shortness of breath, dizzy spells) for 1,000 patients associated with 110 cardiac rehabilitation facilities in a large state over a six-month period. Patients who had recently suffered from a heart attack and who were entering rehabilitation agreed to be randomly assigned to either a new exercise treatment program, or the standard treatment protocol. Of particular interest to the researcher heading up this study is the relationship between treatment condition and the number of cardiac warning incidents. The new approach to rehabilitation is expected to result in fewer such incidents as compared to the traditional method. In addition, the researcher has also collected data on the sex of the patients, and the number of hours that each rehabilitation facility is open during the week. This latter variable is of interest as it reflects the overall availability of the rehabilitation programs. The new method of conducting cardiac rehabilitation is coded in the data as 1, while the standard approach is coded as 0. Males are also coded as 1 while females are assigned a value of 0.

Random Intercept Poisson Regression

The R commands and resultant output for fitting the Poisson regression model to the data appear below using the glmer function in the lme4 library that was employed earlier to fit the dichotomous logistic regression models.

```
summary(model8.7<-glmer(heart~trt+sex+(1|rehab),family=pois
son, data=rehab_data))

Generalized linear mixed model fit by maximum likelihood
(Laplace Approximation) ['glmerMod']
 Family: poisson  ( log )
Formula: heart ~ trt + sex + (1 | rehab)
   Data: rehab_data

     AIC       BIC    logLik deviance df.resid
 11470.2   11489.8   -5731.1   11462.2      996

Scaled residuals:
    Min      1Q Median      3Q     Max
 -5.906  -1.695 -0.881   0.756 40.163

Random effects:
 Groups Name         Variance Std.Dev.
 rehab  (Intercept)  1.216    1.103
Number of obs: 1000, groups:  rehab, 110

Fixed effects:
            Estimate Std. Error z value Pr(>|z|)
(Intercept)  0.83408    0.11181    7.46 8.67e-14 ***
trt         -0.45612    0.03389  -13.46  < 2e-16 ***
sex          0.39305    0.03344   11.76  < 2e-16 ***
---
Signif. codes:  0 '***' 0.001 '**' 0.01 '*' 0.05 '.' 0.1 ' ' 1
```

```
Correlation of Fixed Effects:
    (Intr) trt
trt -0.112
sex -0.167 -0.055
```

In terms of the function call, the syntax for Model 8.7 is virtually identical to that used for the dichotomous logistic regression model. The dependent and independent variables are linked in the usual way that we have seen in R: heart~trt+sex. Here, the outcome variable is heart, which reflects the frequency of the warning signs for heart problems that we described above. The independent variables are treatment (trt) and sex of the individual, while the specific rehabilitation facility is contained in the variable rehab. In this model, we are fitting a random intercept-only, with no random slope and no rehabilitation-center-level variables.

The results of the analysis indicate that there is variation among the intercepts from rehabilitation facility to rehabilitation facility, with a variance of 1.216. As a reminder, the intercept reflects the mean frequency of events when (in this case) both of the independent variables are 0; i.e. females in the control condition. The average intercept across the 110 rehabilitation centers is 1.216, and this non-zero value suggests that the intercept does differ from center to center. Put another way, we can conclude that the mean number of cardiac warning signs varies across rehabilitation centers, and that the average female in the control condition will have approximately 1.2 such incidents over the course of six months. In addition, these results reveal a statistically significant negative relationship between heart and trt, and a statistically significant positive relationship between heart and sex. Remember that the new treatment is coded as 1 and the control as 0, so that a negative relationship indicates that there are fewer warning signs over six months for those in the treatment than those in the control group. Also, given that males were coded as 1 and females as 0, the positive slope for sex means that males have more warning signs on average than do females.

Random Coefficient Poisson Regression

If we believe that the treatment will have different impacts on the number of warning signs present among the rehabilitation centers, we would want to fit the random coefficient model. This can be done for Poisson regression just as it was syntactically for dichotomous logistic regression, as demonstrated in Model 8.8.

```
summary(model8.8<-glmer(heart~trt+sex+(trt|rehab),family=pois
son, data=rehab_data))

Generalized linear mixed model fit by maximum likelihood
(Laplace Approximation) ['glmerMod']
```

```
Family: poisson  ( log )
Formula: heart ~ trt + sex + (trt | rehab)
   Data: rehab_data

     AIC       BIC    logLik  deviance  df.resid
 10554.6  10584.0   -5271.3   10542.6       994

Scaled residuals:
    Min       1Q  Median        3Q      Max
 -6.109  -1.552  -0.725     0.640   31.917

Random effects:
 Groups  Name           Variance  Std.Dev.  Corr
 rehab   (Intercept)    1.869     1.367
         trt            1.844     1.358     -0.62
Number of obs: 1000, groups:  rehab, 110

Fixed effects:
              Estimate  Std. Error  z value  Pr(>|z|)
(Intercept)    0.52852     0.14124    3.742  0.000183 ***
trt           -0.12222     0.14749   -0.829  0.407310
sex            0.34415     0.03523    9.769   < 2e-16 ***
---
Signif. codes:  0 '***' 0.001 '**' 0.01 '*' 0.05 '.' 0.1 ' ' 1

Correlation of Fixed Effects:
     (Intr)  trt
trt  -0.622
sex  -0.137  -0.004
```

The syntax for the inclusion of random slopes in the model is identical to that used with logistic regression and thus will not be commented on further here. The random effect for slopes across rehabilitation centers was estimated to be 1.844, which indicates that there is some differential center effect to the impact of treatment on the number of cardiac warning signs experienced by patients. Indeed, the variance for the random slopes is approximately the same magnitude as the variance for the random intercepts, which indicates that these two random effects are quite comparable in magnitude. The correlation of the random slope and intercept model components is fairly large and negative (–0.62), meaning that the greater the number of cardiac events in a rehab center, the lower the impact of the treatment on the number of such events. The average slope for treatment across centers was no longer statistically significant, which indicates that when we account for the random coefficient effect for treatment, the treatment effect itself goes away.

As with the logistic regression, we can compare the fit of the two models using both information indices, and a likelihood ratio test.

```
anova(model8.7,model8.8)

Data: rehab_data
Models:
model8.7: heart ~ trt + sex + (1 | rehab)
model8.8: heart ~ trt + sex + (trt | rehab)
         Df   AIC    BIC   logLik deviance  Chisq Chi Df
Pr(>Chisq)
model8.7  4 11470 11490 -5731.1    11462
model8.8  6 10555 10584 -5271.3    10543 919.63       2  <
2.2e-16 ***
---
Signif. codes:  0 '***' 0.001 '**' 0.01 '*' 0.05 '.' 0.1 ' ' 1
```

Given that there is a statistically significant difference in model fit ($p < 0.001$), and Model 8.8 has the smaller AIC and BIC values, these results provide further statistical evidence that the relationship of treatment with the number of cardiac symptoms differs across rehabilitation centers.

Inclusion of Additional Level-2 Effects to the Multilevel Poisson Regression Model

You may recall that in addition to testing for treatment and gender differences in the rate of heart warning signs, the researcher conducting this study also wanted to know whether the number of hours per week the rehabilitation centers were open (hours) was related to the outcome variable. In order to address this question, we will need to fit a model with both level-1 (trt and sex) and level-2 (hours) effects.

```
summary(model8.9<-glmer(heart~trt+sex+hours+(1|rehab),family=p
oisson, data=rehab_data))

Generalized linear mixed model fit by maximum likelihood
(Laplace Approximation) ['glmerMod']
 Family: poisson  ( log )
Formula: heart ~ trt + sex + hours + (1 | rehab)
   Data: rehab_data

    AIC      BIC   logLik deviance df.resid
 11467.5  11492.0  -5728.7  11457.5      995

Scaled residuals:
   Min     1Q Median     3Q    Max
-5.907 -1.688 -0.887  0.755 40.150

Random effects:
 Groups Name         Variance Std.Dev.
 rehab  (Intercept) 1.148    1.071
Number of obs: 1000, groups:  rehab, 110
```

```
Fixed effects:
              Estimate Std. Error z value Pr(>|z|)
(Intercept)    0.80895    0.10979   7.368 1.73e-13 ***
trt           -0.45617    0.03389 -13.459  < 2e-16 ***
sex            0.39319    0.03344  11.758  < 2e-16 ***
hours          0.24530    0.11142   2.201   0.0277 *
---
Signif. codes:  0 '***' 0.001 '**' 0.01 '*' 0.05 '.' 0.1 ' ' 1

Correlation of Fixed Effects:
      (Intr) trt    sex
trt   -0.114
sex   -0.171 -0.056
hours -0.123  0.001  0.002
```

These results show that the more hours a center is open the more warning signs patients who attend will experience over a six-month period. In other respects, the parameter estimates for model8.9 do not differ substantially from those of the earlier models, generally revealing similar relationships among the independent and dependent variables.

We can also make comparisons among the various models in order to determine which yields the best fit to our data. Given that the AIC and BIC values for model8.8 are lower than those of model8.9, we would conclude that model8.8 yields the best fit to the data. In addition, below are the results for the likelihood ratio tests comparing these models with one another.

```
anova(model8.8,model8.9)

Data: rehab_data
Models:
model8.9: heart ~ trt + sex + hours + (1 | rehab)
model8.8: heart ~ trt + sex + (trt | rehab)
         Df   AIC   BIC  logLik deviance  Chisq Chi Df
Pr(>Chisq)
model8.9  5 11468 11492 -5728.7    11458
model8.8  6 10555 10584 -5271.3    10543 914.95      1  <
2.2e-16 ***
---
Signif. codes:  0 '***' 0.001 '**' 0.01 '*' 0.05 '.' 0.1 ' ' 1

anova(model8.7,model8.9)

Data: rehab_data
Models:
model8.7: heart ~ trt + sex + (1 | rehab)
model8.9: heart ~ trt + sex + hours + (1 | rehab)
         Df   AIC   BIC  logLik deviance  Chisq Chi Df
Pr(>Chisq)
```

```
model8.7   4  11470  11490  -5731.1      11462
model8.9   5  11468  11492  -5728.7      11458  4.6896        1
0.03035 *
---
Signif. codes:  0 `***' 0.001 `**' 0.01 `*' 0.05 `.' 0.1 ` ' 1
```

These results show that model8.8 fits the data significantly better than
model8.9, which in turn fits the data significantly better than model8.7.
Earlier, we found that model8.8 also fit the data better than model8.7, based
on the likelihood ratio test results, and AIC/BIC values. Thus, given all of
these results, we would conclude that model8.8 provides the best fit to the
data, from among the three that we tried here.

Recall that the signal quality of the Poisson distribution is the equality
of the mean and variance. In some instances, however, the variance of
a variable may be larger than the mean, leading to the problem of overdispersion, which we described in Chapter 7. In the previous chapter we
described alternative statistical models for such situations, including
one based on the quasipoisson distribution, which took the same form
as the Poisson, except that it relaxed the requirement of equal mean and
variance. It is possible to fit the quasipoisson distribution in the multilevel modeling context as well, though not using lme4. The developer of
lme4 is not confident in the quasipoisson fitting algorithm, and has thus
removed this functionality from lme4, though alternative estimators for
overdispersed data are available using lme4. Rather, we would need to
use the glmmPQL package from the nlme library. In this case, we would
use the following syntax for the random intercept model with the quasipoisson estimator.

```
summary(model8.10<-glmmPQL(heart~trt+sex,random=~1|rehab,famil
y=quasipoisson))

Linear mixed-effects model fit by maximum likelihood
 Data: NULL
  AIC BIC logLik
   NA  NA     NA

Random effects:
 Formula: ~1 | rehab
         (Intercept) Residual
StdDev:   0.6620581 4.010266

Variance function:
 Structure: fixed weights
 Formula: ~invwt
Fixed effects: heart ~ trt + sex
                Value  Std.Error  DF   t-value  p-value
(Intercept) 1.2306419 0.09601707 888 12.816908   0.0000
```

```
trt          -0.2163649 0.06482216 888 -3.337823  0.0009
sex           0.1354837 0.06305874 888  2.148531  0.0319
 Correlation:
     (Intr) trt
trt -0.152
sex -0.099  0.045
```

```
Standardized Within-Group Residuals:
        Min          Q1         Med          Q3         Max
-1.48542222 -0.46850088 -0.36061041  0.09957005 12.49933170
```

```
Number of Observations: 1000
Number of Groups: 110
```

The results of the quasipoisson regression are essentially identical to those in Model 8.10, which used Poisson regression. Thus, it would appear that there is not a problem with overdispersion in the data. In this instance, we can rely on the Poisson regression results with some confidence.

As we learned in the previous chapter, the negative binomial distribution presents another alternative for use when the outcome variable is overdispersed. Unlike quasipoisson regression, in which the distribution is essentially Poisson with a relaxation on the requirement that $\phi = 1$, the negative binomial distribution takes an alternate form from the Poisson, with a difference in the variance parameter (see Chapter 7 for a discussion of this difference). In order to fit the negative binomial model, we will use the glmer.nb function in the glmer library.

```
summary(model8.11<-glmer.nb(heart~trt+sex+(1|rehab),
data=rehab_data))
```

```
Generalized linear mixed model fit by maximum likelihood
(Laplace Approximation) ['glmerMod']
 Family: Negative Binomial(0.1825)  ( log )
Formula: heart ~ trt + sex + (1 | rehab)
   Data: rehab_data

     AIC      BIC   logLik deviance df.resid
  3937.9   3962.5  -1964.0   3927.9      995

Scaled residuals:
    Min      1Q  Median      3Q     Max
-0.4240 -0.4170 -0.4101  0.1304  9.6989

Random effects:
 Groups Name        Variance Std.Dev.
 rehab  (Intercept) 0.2079   0.456
Number of obs: 1000, groups:  rehab, 110
```

```
Fixed effects:
             Estimate  Std. Error  z value  Pr(>|z|)
(Intercept)    1.2140      0.1541    7.879   3.3e-15 ***
trt           -0.5126      0.1708   -3.001   0.00269 **
sex            0.4189      0.1729    2.423   0.01541 *
---
Signif. codes:  0 '***' 0.001 '**' 0.01 '*' 0.05 '.' 0.1 ' ' 1

Correlation of Fixed Effects:
    (Intr) trt
trt -0.535
sex -0.282 -0.271
```

The function call includes the standard model setup in R for the fixed effects (trt, sex), with the random effect (intercept within school in this example) denoted as for the glmer-based models. In terms of the output, after the function call, we see the table of parameter estimates, standard errors, test statistics, and *p*-values. These results are similar to those described above, indicating the significant relationships between the frequency of cardiac warning signs and both treatment and sex. The variance associated with the random effect was estimated to be 0.2079. The findings with respect to the fixed effects are essentially the same as those for the standard Poisson regression model, with a statistically significant negative relationship between treatment and the number of cardiac events, and a significant positive relationship for sex. The 95% profile confidence intervals for the fixed and random effects in model8.11 appear below.

```
confint(model8.11, method=c("profile"))

Computing profile confidence intervals ...
                  2.5 %        97.5 %
.sig01        0.17089277    0.7164493
(Intercept)   0.90993733    1.5155525
trt          -0.84534793   -0.1752563
sex           0.07584926    0.7545013
```

From these results, we can see that 0 is not included in any of the intervals, meaning that they are all statistically significant.

As with the Poisson regression model, it is possible to fit a random coefficients model for the negative binomial, using very similar R syntax as that for glmer. In this case, we will fit a random coefficient for the trt variable, as we did for the Poisson regression model.

```
summary(model8.12<-glmer.nb(heart~trt+sex+(trt|rehab),
data=rehab_data))

Generalized linear mixed model fit by maximum likelihood
(Laplace Approximation) ['glmerMod']
```

```
Family: Negative Binomial(0.1845)  ( log )
Formula: heart ~ trt + sex + (trt | rehab)
   Data: rehab_data

     AIC      BIC   logLik deviance df.resid
  3939.3   3973.7  -1962.7   3925.3      993

Scaled residuals:
    Min      1Q  Median      3Q     Max
-0.4267 -0.4180 -0.4121  0.1375  8.0981

Random effects:
 Groups Name        Variance Std.Dev. Corr
 rehab  (Intercept) 0.36992  0.6082
        trt         0.09906  0.3147   -1.00
Number of obs: 1000, groups:  rehab, 110

Fixed effects:
            Estimate Std. Error z value Pr(>|z|)
(Intercept)   1.1155     0.1694   6.583 4.61e-11 ***
trt          -0.3487     0.1986  -1.756   0.0791 .
sex           0.4000     0.1717   2.330   0.0198 *
---
Signif. codes:  0 '***' 0.001 '**' 0.01 '*' 0.05 '.' 0.1 ' ' 1

Correlation of Fixed Effects:
    (Intr) trt
trt -0.643
sex -0.246 -0.276
```

The variance estimate for the random `trt` effect is 0.09906, as compared to the larger random intercept variance estimate of 0.36992. This result suggests that the differences in the mean number of cardiac events across rehab centers is greater than the cross-center differences in the treatment effect on the number of events. The 95% percentile bootstrap confidence intervals for the fixed and random effects appear below. Note that the profile confidence interval approach did not converge, and thus wasn't used here.

```
confint(model8.12, method=c("boot"), boot.type=c("perc"))

Computing bootstrap confidence intervals ...
                  2.5 %      97.5 %
.sig01       0.13474208  0.81403061
.sig02      -1.00000000  0.99999999
.sig03       0.01219854  0.85568955
(Intercept)  0.66224672  1.40871914
trt         -0.68082872  0.03825103
sex          0.07316333  0.68937019
```

The intervals for the random intercept (.sig01), the random coefficient (.sig03), the fixed intercept, and sex all excluded 0, meaning that these terms can be seen as statistically significant. However, intervals for the correlation between the random effects (.sig02) and the fixed treatment effect (trt) all included 0. Thus, we would conclude that there is not a statistically significant relationship between treatment condition and the number of cardiac events, when the random treatment effect is included in the model. This conclusion based on the confidence interval mirrors the result of the hypothesis test in the original output for model8.12. There is, however, a difference in the treatment effect across the rehab centers, given the statistically significant random coefficient effect.

As with other models that we have examined in this book, it is possible to include level-2 independent variables, such as number of hours the centers are open, in the model, and compare the relative fit using the relative fit indices, as in Model8.13.

```
summary(model8.13<-glmer.nb(heart~trt+sex+hours+(1|rehab),
data=rehab_data))

Generalized linear mixed model fit by maximum likelihood
(Laplace Approximation) ['glmerMod']
 Family: Negative Binomial(0.1813)  ( log )
Formula: heart ~ trt + sex + hours + (1 | rehab)
   Data: rehab_data

    AIC      BIC   logLik deviance df.resid
 3932.8   3962.2  -1960.4   3920.8      994

Scaled residuals:
   Min      1Q  Median      3Q     Max
-0.4223 -0.4161 -0.4090  0.1177  9.4494

Random effects:
 Groups Name         Variance Std.Dev.
 rehab  (Intercept) 0.1284   0.3583
Number of obs: 1000, groups:  rehab, 110

Fixed effects:
            Estimate Std. Error z value Pr(>|z|)
(Intercept)  1.20520    0.15109   7.977  1.5e-15 ***
trt         -0.49358    0.16690  -2.957  0.00310 **
sex          0.40764    0.16925   2.408  0.01602 *
hours        0.26415    0.09546   2.767  0.00566 **
---
Signif. codes:  0 '***' 0.001 '**' 0.01 '*' 0.05 '.' 0.1 ' ' 1

Correlation of Fixed Effects:
      (Intr) trt    sex
```

```
trt    -0.534
sex    -0.292 -0.246
hours  -0.066  0.104 -0.082
```

As we have seen previously, the number of hours that the centers are open is significantly positively related to the number of cardiac warning signs over the six-month period of the study. The 95% profile confidence intervals for model8.13 appear below.

The 95% profile confidence intervals for the random and fixed effects in model8.13 appear below. The interval for the random intercept effect (.sig01) includes 0, leading us to conclude that under the negative binomial model when the number of hours for which the center is open is included in the model, there is not a statistically significant difference among the rehab centers in terms of the number of cardiac events.

```
confint(model8.13, method=c("profile"))

Computing profile confidence intervals ...
                    2.5 %        97.5 %
.sig01         0.00000000    0.6353885
(Intercept)    0.90671458    1.5005578
trt           -0.81897539   -0.1638816
sex            0.07145093    0.7359959
hours          0.07281699    0.4504233
```

Now that we have examined the random and fixed effects for each of the three negative binomial models, we can compare their fit with one another in order to determine which is optimal given the set of data at hand. We can formally compare these three models using the likelihood ratio test, as below. In addition, we will also compare the AIC and BIC values to provide further evidence regarding the relative model fit.

```
anova(model8.11, model8.12)

Data: rehab_data
Models:
model8.11: heart ~ trt + sex + (1 | rehab)
model8.12: heart ~ trt + sex + (trt | rehab)
          Df    AIC     BIC   logLik deviance  Chisq Chi Df
Pr(>Chisq)
model8.11  5 3937.9 3962.5 -1964.0   3927.9
model8.12  7 3939.3 3973.7 -1962.7   3925.3 2.6163      2
0.2703

anova(model8.11, model8.13)

Data: rehab_data
Models:
```

```
model8.11: heart ~ trt + sex + (1 | rehab)
model8.13: heart ~ trt + sex + hours + (1 | rehab)
         Df    AIC    BIC  logLik deviance  Chisq Chi Df
Pr(>Chisq)
model8.11  5 3937.9 3962.5 -1964.0   3927.9
model8.13  6 3932.8 3962.2 -1960.4   3920.8 7.1424      1
0.007528 **
---
Signif. codes:  0 '***' 0.001 '**' 0.01 '*' 0.05 '.' 0.1 ' ' 1
```

First, the results of the likelihood ratio test show that the fit of Model8.12 was not significantly different than that of Model8.11, meaning that inclusion of the random treatment effect did not improve model fit. This finding is further reinforced by the lower AIC and BIC values for model8.11. Next, we compared the fit of models8.11 and 8.13. Here, we see a statistically significant difference in the model fit, with slightly lower AIC and BIC values associated with model8.13, which included the number of hours that the rehab centers were open. Thus, we would conclude that having this variable in the model is associated with better fit to the data. Finally, models8.12 and 8.13 are not nested within one another, and thus cannot be compared using the likelihood ratio test. However, the AIC and BIC values for model8.13 were smaller than those of model8.12, suggesting that inclusion of the rehab center hours is more important to yielding good fit to the data than is inclusion of the random treatment condition effect.

Summary

In Chapter 8 we learned that the generalized linear models featured in Chapter 7, which accommodate categorical dependent variables, can be easily extended to the multilevel context. Indeed, the basic concepts that we learned in Chapter 2 regarding sources of variation, and various types of models can be easily extended for categorical outcomes. In addition, R provides for easy fitting of such models through the lme and lmer families of functions. Therefore, in many ways Chapter 8 represents a review of material that by now should be familiar to us, even while applied in a different scenario than we have seen up to now. Perhaps the most important point to take away from this chapter is the notion that modeling multilevel data in the context of generalized linear models is not radically different from the normally distributed continuous dependent variable case, so that the same types of interpretations can be made, and the same types of data structure can be accommodated.

9

Bayesian Multilevel Modeling

Bayesian statistical modeling represents a fundamental shift from the frequentist methods of model parameter estimation that we have been using heretofore. This paradigm shift is evident in part through the methodology used to obtain the estimates, Markov Chain Monte Carlo (MCMC) most commonly for the Bayesian approach, and maximum likelihood (ML) and restricted maximum likelihood (REML) in the frequentist case. In addition, Bayesian estimation involves the use of prior distributional information that is not present in frequentist-based approaches. Perhaps even more than the obvious methodological differences, however, the Bayesian analytic framework involves a very different view from that traditionally espoused in the likelihood-based literature as to the nature of population parameters. In particular, frequentist-based methods estimate the population parameter using a single value that is obtained using the sample data only. In contrast, in the Bayesian paradigm the population parameter is estimated as a distribution of values, rather than a single number. Furthermore, this estimation is carried out using both the sample data and prior distribution information provided by the researcher. Bayesian methods combine this prior information regarding the nature of the parameter distribution with information taken from the sample data in order to estimate a posterior distribution of the parameter distribution. In practice, when a single value estimate of a model parameter is desired, such as a regression coefficient-linking dependent and independent variables, the mean, median, or mode of the posterior distribution is calculated. In addition, standard deviations and density intervals for model parameters can also be estimated from this posterior distribution.

A key component of conducting Bayesian analysis is the specification of a prior distribution for each of the model parameters. These priors can be either one of two types. Informative priors are typically drawn from prior research, and will be fairly specific in terms of both their mean and variance. For example, a researcher may find a number of studies in which a vocabulary test score has been used to predict reading achievement. Perhaps across these studies the regression coefficient is consistently around 0.5. The researcher may then set the prior for this coefficient as the normal distribution with a mean of 0.5 and a variance of 0.1. In doing so, (s)he is stating up front that the coefficient linking these two variables in their own study is likely to be near this value. Of course, such may not be the case, and because the data are also used to obtain the posterior distribution, the prior plays only a partial role in its determination. In contrast to informative priors, noninformative (sometimes referred to as diffuse) priors are not based on prior research. Rather, noninformative priors are deliberately selected so as to constrain the posterior distribution for the parameter as little as possible, in light of the fact that little or no useful information is available for setting the prior distribution. As an example, if there is not sufficient evidence in the literature for the researcher to know what the distribution of the regression coefficient is likely to be, (s)he may set the prior as normal with a mean of 0 and a large variance of, perhaps, 1,000 or even more. By using such a large variance for the prior distribution, the researcher is acknowledging the lack of credible information regarding what the posterior distribution might be, thereby leaving the posterior distribution largely unaffected by the prior, and relying primarily on the observed data to obtain the parameter estimate.

The reader may rightly question why, or in what situations, Bayesian multilevel modeling might be particularly useful, or even preferable to frequentist methods. One primary advantage of Bayesian methods in some situations, including with multilevel modeling, is that unlike ML and REML, it does not rely on any distributional assumptions about the data. Thus, the determination of Bayesian credibility intervals (corresponding to confidence intervals) can be made without worry even if the data come from a skewed distribution. In contrast, ML or REML confidence intervals may not be accurate if foundational distributional assumptions are not met. In addition, the Bayesian approach can be quite useful when the model to be estimated is very complex and frequentist-based approaches such as ML and REML are not able to converge. A related advantage is that the Bayesian approach may be better able to provide accurate model parameter estimates in the small sample case. And, of course, the Bayesian approach to parameter estimation can be used in cases where ML and REML also work well, and as we will see below, the different methods generally yield similar results in such situations.

The scope of this book does not encompass the technical aspects of MCMC estimation, which is most commonly used to obtain Bayesian estimates. The interested reader is encouraged to reference any of several good works on the topic. In particular, Lynch (2010) provides a very thorough introduction to Bayesian methods for social scientists, including a discussion of the MCMC algorithm, and Kruschke (2011) provides a nice general description of applied Bayesian analysis. It should be noted here that while MCMC is the most frequently used approach for parameter estimation in the Bayesian context, it is not itself inherently Bayesian. Rather, it is simply an algorithmic approach to sampling from complex sampling distributions, such as a posterior distribution that might be seen with complicated models such as those in the multilevel context. Although we will not describe the MCMC process in much detail here, it is necessary to discuss conceptually how the methodology works, in part so that you, the reader, might be more comfortable with where the parameter estimates come from, and in part because we will need to diagnose whether the method has worked appropriately so that we can have confidence in the final parameter estimates.

MCMC is an iterative process in which the prior distribution is combined with information from the actual sample in order to obtain an estimate of the posterior distributions for each of the model parameters (e.g. regression coefficients, random-effect variances). From this posterior distribution, parameter values are simulated a large number of times in order to obtain an estimated posterior distribution. After each such sample is drawn, the posterior is updated. This iterative sampling and updating process is repeated a very large number of times (e.g. 10,000 or more) until there is evidence of convergence regarding the posterior distribution; i.e. a value from one sampling draw is very similar to the previous sample draw. The Markov Chain part of MCMC reflects the process of sampling a current value from the posterior distribution, given the previous sampled value, while Monte Carlo reflects the random simulation of these values from the posterior distribution. When the chain of values has converged, we are left with an estimate of the posterior distribution of the parameter of interest (e.g. regression coefficient). At this point, a single model parameter estimate can be obtained by calculating the mean, median, or mode from the posterior distribution.

When using MCMC, the researcher must be aware of some technical aspects of the estimation that need to be assessed in order to ensure that the analysis has worked properly. The collection of 10,000 (or more) individual parameter estimates forms a lengthy time series, which must be examined to ensure that two things are true. First, the parameter estimates must converge, and second the autocorrelation between different iterations in the process should be low. Parameter convergence can be assessed through the use of a trace plot, which is simply a graph of the parameter estimates in order from the first iteration to the last. The autocorrelation of estimates is calculated for a variety of iterations, and the researcher will look for the distance between estimates at which the autocorrelation becomes quite low. When it is determined at what point the autocorrelation between estimates is sufficiently low, the estimates are thinned so as to remove those that might be more highly autocorrelated with one another than would be desirable. So, for example, if the autocorrelation is low when the estimates are ten iterations apart, the time series of 10,000 sample points would be thinned to include only every tenth observation, in order to create the posterior distribution of the parameter. The mean/median/mode of this distribution would then be calculated using only the thinned values, in order to obtain the single parameter estimate value that is reported by R. A final issue in this regard is what is known as the burn-in period. Thinking back to the issue of distributional convergence, the researcher will not want to include any values in the posterior distribution for iterations prior to the point at which the time series converged. Thus, iterations prior to this convergence are referred to as having occurred during the burn-in period, and are not used in the calculation of posterior means/medians/modes. Each of these MCMC conditions (number of iterations, thinning rate, and burn-in period) can be set by the user in R, or default values can be used. In the remainder of this chapter we will provide detailed examples of the diagnosis of MCMC results, and the setting not only of MCMC parameters, but also of prior distributions.

MCMCglmm for a Normally Distributed Response Variable

We will begin our discussion of fitting a random intercept model with the Prime Time data, which we have used in numerous examples in previous chapters. In particular, we will fit a model in which reading achievement score is the dependent variable, and vocabulary score is the independent variable. Students are nested within schools, which we will treat as a random effect. Bayesian multilevel modeling can be done in R using the MCMCglmm library. As we discussed previously in this chapter, a key component of Bayesian modeling is the use of prior distribution information in the estimation of the posterior distribution of the model parameters. MCMCglmm has a default set of priors that it uses for each model parameter, and upon which we will rely for the first example analysis. The default priors for the model coefficients and intercepts are noninformative in nature, taken from the standard normal distribution with a mean of 0, and a variance of 1e + 10, or 10,000,000,000. This very large variance for the prior reflects our relative lack of confidence that the mean of the coefficient distributions is in fact 0. The prior distribution for the random effect, school, is referred to in MCMCglmm parlance as the G-structure, and is expressed using two separate terms: (1) V reflecting the variation in the outcome variable (reading score) across schools, and (2) nu reflecting the degree of belief in the parameter. The default prior distribution for V is the inverse-Wishart distribution, with V = 1 and nu = 0. This low value for nu reflects the lack of information provided by the prior distribution. There is also a prior distribution for the residual term, R, with the defaults being precisely the same as those for G.

In order to fit the random intercept model with a single predictor under the Bayesian framework with default priors, we will use the following commands in R:

```
library(MCMCglmm)
prime_time.nomiss<-na.omit(prime_time)
attach(prime_time.nomiss)
model9.1<-MCMCglmm(geread~gevocab, random=~school, data=prime_
time.nomiss)
plot(model9.1)
summary(model9.1)
```

The function call for MCMCglmm is fairly similar to what we have seen in previous chapters. One important point to note is that MCMCglmm does not accommodate the presence of missing data. Therefore, before conducting the analysis we needed to expunge all of the observations with missing data. We created a dataset with no missing observations using the command prime _ time.nomiss<-na.omit(prime _ time), which created a new data frame called prime _ time.nomiss containing no missing data. We then attached this data frame and fit the multilevel model, indicating the random effect

with the `random=~` statement. We subsequently requested a summary of
the results and a plot of the relevant graphs that will be used in determining
whether the Bayesian model has converged properly. It is important to note
that by default, `MCMCglmm` uses 13,000 iterations of the MCMC algorithm,
with a burn-in of 3,000 and thinning of 10. As we will see below, we can easily adjust these settings to best suit our specific analysis problem.

When interpreting the results of the Bayesian analysis, we first want to
know whether we can be confident in the quality of the parameter estimates
for both the fixed and random effects. The plots relevant to this diagnosis
appear below.

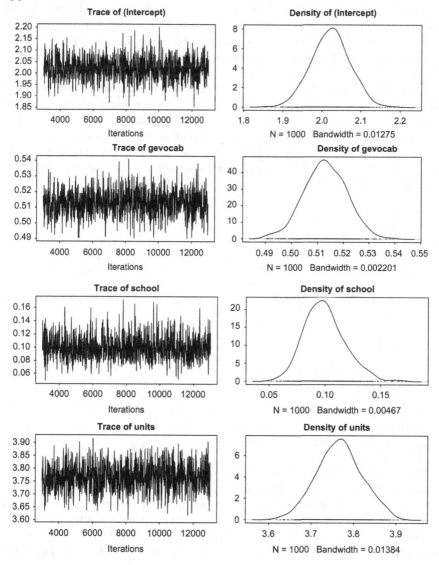

For each model parameter, we have the trace plot on the left, showing the entire set of parameter estimates as a time series across the 13,000 iterations. On the right, we have a histogram of the distribution of parameter estimates. Our purpose for examining these plots is to ascertain to what extent the estimates have converged on a single value. As an example, the first pair of graphs reflects the parameter estimates for the intercept. For the trace, convergence is indicated when the time series plot hovers around a single value on the y-axis, and does not meander up and down. In this case, it is clear that the trace plot for the intercept shows convergence. This conclusion is reinforced by the histogram for the estimate, which is clearly centered over a single mean value, with no bimodal tendencies. We see similar results for the coefficient of vocabulary, the random effect of school, and the residual. Given that the parameter estimates appear to have successfully converged, we can have confidence in the actual estimated values, which we will examine shortly.

Prior to looking at the parameter estimates, we want to assess the autocorrelation of the estimates in the time series for each parameter. Our purpose here is to ensure that the rate of thinning (taking every tenth observation generated by the MCMC algorithm) that we used is sufficient to ensure that any autocorrelation in the estimates is eliminated. In order to obtain the autocorrelations for the random effects, we use the command autocorr(model9.1$VCV), and obtain the following results.

```
, , school

              school          units
Lag 0      1.00000000  -0.05486644
Lag 10    -0.03926722  -0.03504799
Lag 50    -0.01636431  -0.04016879
Lag 100   -0.03545104   0.01987726
Lag 500    0.04274662  -0.05083669


, , units

                school           units
Lag 0    -0.0548664421   1.000000000
Lag 10   -0.0280445140  -0.006663408
Lag 50   -0.0098424151   0.017031804
Lag 100   0.0002654196   0.010154987
Lag 500  -0.0022835508   0.046769152
```

We read this table as follows: in the first section, we see results for the random effect school. This output includes correlations involving the school variance component estimates. Under the school column are the actual autocorrelations for the school random effect estimate. Under the units column are the cross correlations between estimates for the school random effect and the residual random effect, at different lags. Thus, for example, the correlation between the estimates for school and the residual with no lag is −0.0549. The correlation between the school estimate 10 lags prior to the current residual estimate is −0.035. In terms of ascertaining whether our rate of thinning is sufficient, the more important numbers are in the school column, where we see the correlation between a given school effect estimate and the school effect estimate 10, 50, 100, and 500 estimates before. The auto-correlation at a lag value of 10, −0.0393, is sufficiently small for us to have confidence in our thinning the results at 10. We would reach a similar conclusion regarding the autocorrelation of the residual (units), such that 10 appears to be a reasonable thinning value for it as well. We can obtain the autocorre-lations of the fixed effects using the command autocorr(model9.1$Sol). Once again, it is clear that there is essentially no autocorrelation as far out as a lag of 10, indicating that the default thinning value of 10 is sufficient for both the intercept and the vocabulary test score.

```
, , (Intercept)

          (Intercept)         gevocab
Lag 0      1.000000000  -0.757915532
Lag 10    -0.002544175  -0.013266125
Lag 50    -0.019405970   0.007370979
Lag 100   -0.054852949   0.029253018
Lag 500    0.065853783  -0.046153346

, , gevocab

          (Intercept)         gevocab
Lag 0     -0.757915532   1.000000000
Lag 10     0.008583659   0.020942660
Lag 50    -0.001197203  -0.002538901
Lag 100    0.047596351  -0.022549594
Lag 500   -0.057219532   0.026075911
```

Having established that the parameter estimates have converged properly, and that our rate of thinning in the sampling of MCMCderived values is suf-ficient to eliminate any autocorrelation in the estimates, we are now ready to examine the specific parameter estimates for our model. The output for this analysis appears below.

```
Iterations = 3001:12991
 Thinning interval  = 10
 Sample size  = 1000

DIC: 43074.14

G-structure:   ~school

        post.mean 1-95% CI u-95% CI eff.samp
```

```
school    0.09962   0.06991    0.1419       1000
```

```
 R-structure:  ~units
```

```
        post.mean 1-95% CI u-95% CI eff.samp
units       3.767    3.668     3.876       1000
```

```
 Location effects: geread ~ gevocab
```

```
             post.mean 1-95% CI u-95% CI eff.samp  pMCMC
(Intercept)     2.0220   1.9323    2.1232      1000 <0.001 ***
gevocab         0.5131   0.4975    0.5299      1000 <0.001 ***
---
Signif. codes:  0 '***' 0.001 '**' 0.01 '*' 0.05 '.' 0.1 ' ' 1
```

We are first given information about the number of iterations, the thinning interval, and the final number of MCMC values that were sampled (Sample size) and used to estimate the model parameters. Next, we have the model fit index, the DIC, which can be used for comparing various models and selecting the one that provides optimal fit. The DIC is interpreted in much the same fashion as the AIC and BIC, which we discussed in earlier chapters, and for which smaller values indicate better model fit. We are then provided with the posterior mean of the distribution for each of the random effects, school and residual, which MCMCglmm refers to as units. The mean variance estimate for the school random effect is 0.09962, with a 95% credibility interval of 0.06991 to 0.1419. Remember that we interpret credibility intervals in Bayesian modeling in much the same way that we interpret confidence intervals in frequentist modeling. This result indicates that reading achievement scores do differ across schools, because 0 is not in the interval. Similarly, the residual variance also differs from 0. With regard to the fixed effect of vocabulary score, which had a mean posterior value of 0.5131, we also conclude that the results are statistically significant, given that 0 is not in its 95% credibility interval. We also have a *p*-value for this effect, and the intercept, both of which are significant with values lower than 0.05. The positive value of the posterior mean indicates that students with higher vocabulary scores also had higher reading scores.

In order to demonstrate how we can change the number of iterations, the burn-'in period, and the rate of thinning in R, we will reestimate Model9.1 with 100,000 iterations, a burn-in of 10,000, and a thinning rate of 50. This will yield 1,800 samples for the purposes of estimating the posterior distribution for each model parameter. The R commands for fitting this model, followed by the relevant output, appear below.

```
model9.1b<-MCMCglmm(geread~gevocab, random=~school,
data=prime_time.nomiss, nitt=100000, thin=50, burnin=10000)
plot(model9.1b)
summary(model9.1b)
```

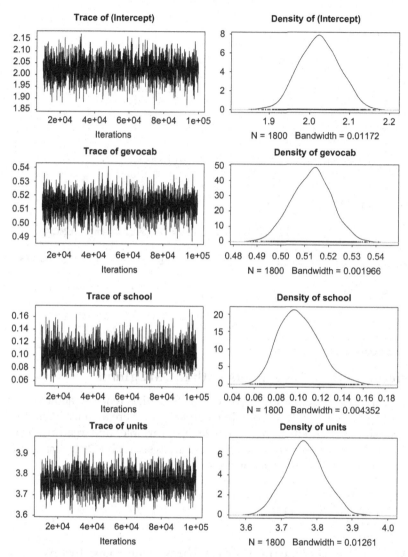

As with the initial model, all parameter estimates appear to have successfully converged. The results, in terms of the posterior means, are also very similar to what we obtained using the default values for the number of iterations, the burn-in period, and the thinning rate. This result is not surprising, given that the diagnostic information for our initial model was all very positive. Nonetheless, it was useful for us to see how the default values can be changed if we need to do so.

```
Iterations = 10001:99951
 Thinning interval  = 50
 Sample size  = 1800
```

```
DIC: 43074.19

G-structure:  ~school

        post.mean l-95% CI u-95% CI eff.samp
school    0.1013   0.06601    0.1366      1800

R-structure:  ~units

        post.mean l-95% CI u-95% CI eff.samp
units     3.766      3.664     3.873      1800

Location effects: geread ~ gevocab

              post.mean l-95% CI u-95% CI eff.samp  pMCMC
(Intercept)      2.0240   1.9304    2.1177     1800 <6e-04 ***
gevocab          0.5128   0.4966    0.5287     1846 <6e-04 ***
---
Signif. codes:  0 '***' 0.001 '**' 0.01 '*' 0.05 '.' 0.1 ' ' 1
```

Including Level-2 Predictors with MCMCglmm

In addition to understanding the extent to which reading achievement is related to vocabulary test score, in Chapter 3 we were also interested in the relationship of school (senroll), a level-2 variable, and reading achievement. Including a level-2 variable in the analysis with MCMCglmm is just as simple as doing so using lme or lme4.

```
model9.2<-MCMCglmm(geread~gevocab+senroll, random=~school,
data=prime_time.nomiss)
plot(model9.2)
```

An examination of the trace plots and histograms shows that we achieved convergence for all parameter estimates. The autocorrelations appear below the graphs, and reveal that the default thinning rate of 10 may not be sufficient so as to remove autocorrelation from the estimates for the intercept and school enrollment. Thus, we refit the model with 40,000 iterations, a burn-in of 3,000, and a thinning rate of 100. We selected a thinning rate of 100 because for each of the model terms, the autocorrelation at a lag of 100, displayed in the results below, was sufficiently small.

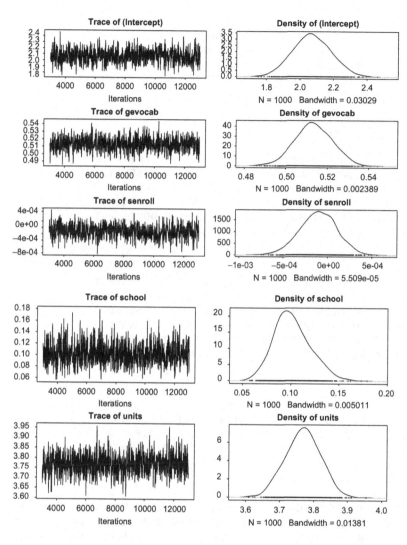

```
autocorr(model9.2$VCV)
, , school

                 school           units
Lag 0      1.000000000  -0.05429139
Lag 10    -0.002457293  -0.07661475
Lag 50    -0.020781555  -0.01761532
Lag 100   -0.027670953   0.01655270
Lag 500    0.035838857  -0.03714127

, , units
```

```
              school            units
Lag 0     -0.05429139   1.000000000
Lag 10     0.03284220  -0.004188523
Lag 50     0.02396060  -0.043733590
Lag 100   -0.04543941  -0.017212479
Lag 500   -0.01812893   0.067148463

autocorr(model9.2$Sol)
, , (Intercept)

              (Intercept)          gevocab          senroll
Lag 0     1.0000000000   -0.3316674622    -0.885551431
Lag 10    0.0801986410   -0.0668713485    -0.064378629
Lag 50    0.0581330411   -0.0348434078    -0.046088343
Lag 100   0.0004512485    0.0001044589    -0.002634201
Lag 500   0.0354993059   -0.0317823452    -0.033329073


, , gevocab

              (Intercept)          gevocab          senroll
Lag 0     -0.331667462   1.0000000000    -0.043353290
Lag 10    -0.014132944   0.0001538528     0.015876989
Lag 50    -0.001177506  -0.0095964368     0.006400198
Lag 100   -0.010782011   0.0143615330     0.004853953
Lag 500   -0.010100604   0.0464368692    -0.017855000


, , senroll

              (Intercept)          gevocab          senroll
Lag 0     -0.8855514315   -0.04335329   1.000000000
Lag 10    -0.0792592927    0.07415593   0.059787652
Lag 50    -0.0542189405    0.04008488   0.037617806
Lag 100    0.0006296859   -0.01189656   0.002608636
Lag 500   -0.0405712255    0.02456735   0.044365323
```

The summary results for the model with 40,000 iterations, and a thinning rate of 100, appear below. It should be noted that the trace plots and histograms of parameter estimates for Model9.2 indicated that convergence had been attained. From these results we can see that the overall fit, based on the DIC, is virtually identical to that of the model not including senroll. In addition, the posterior mean estimate and associated 95% credible interval for this parameter show that senroll was not statistically significantly related to reading achievement; i.e. 0 is in the interval. Taken together, we would conclude that school size does not contribute significantly to the variation in reading achievement scores, nor to the overall fit of the model.

```
summary(model9.2)

 Iterations = 3001:39901
 Thinning interval  = 100
 Sample size  = 1700
```

```
DIC: 43074.86

G-structure:  ~school

        post.mean 1-95% CI u-95% CI eff.samp
school    0.1009   0.06536   0.1373     1000

 R-structure:  ~units

        post.mean 1-95% CI u-95% CI eff.samp
units     3.767    3.664     3.865      1000

 Location effects: geread ~ gevocab + senroll

              post.mean   1-95% CI   u-95% CI eff.samp  pMCMC
(Intercept)   2.072e+00  1.893e+00  2.309e+00    202.2 <0.006
**
gevocab       5.124e-01  4.977e-01  5.282e-01    170.0 <0.006
**
senroll      -9.668e-05 -5.079e-04  3.166e-04    168.3  0.718
---
Signif. codes:  0 '***' 0.001 '**' 0.01 '*' 0.05 '.' 0.1 ' ' 1
```

As a final separate example in this section, we will fit a random coefficients model, in which we allow the relationship of vocabulary score and reading achievement to vary across schools. The syntax for fitting this model with MCMCglmm appears below.

```
model9.3<-MCMCglmm(geread~gevocab, random=~school+gevocab,
data=prime_time.nomiss)
plot(model9.3)
summary(model9.3)
```

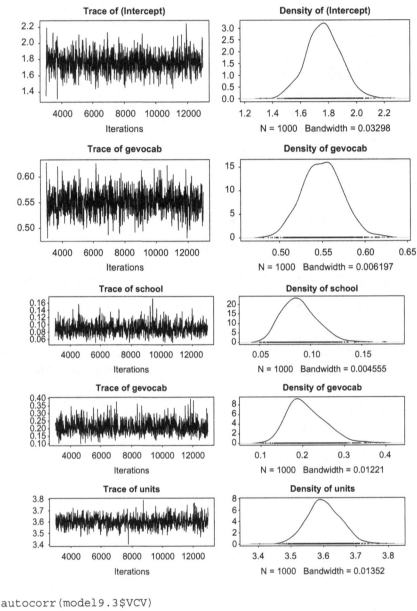

```
autocorr(model9.3$VCV)
, , school

                school        gevocab             units
Lag 0      1.00000000  -0.038280818  -0.054498656
Lag 10     0.03421010   0.019008381   0.003109740
Lag 50    -0.06037994  -0.015998758   0.022603955
Lag 100    0.01134427   0.006434794   0.033359310
Lag 500   -0.01013541  -0.031607061   0.009573277
```

, , gevocab

	school	gevocab	units
Lag 0	-0.038280818	1.00000000	-2.097586e-02
Lag 10	-0.006587315	-0.02620485	4.294747e-02
Lag 50	0.027904335	0.01070891	4.874694e-02
Lag 100	0.082732647	0.03095601	8.865174e-05
Lag 500	0.042865039	-0.03198690	-5.984689e-03

, , units

	school	gevocab	units
Lag 0	-0.05449866	-0.020975858	1.000000000
Lag 10	-0.03789363	0.006081220	0.005303022
Lag 50	0.01538962	-0.006572823	0.004836022
Lag 100	0.01048834	-0.006523078	-0.023194599
Lag 500	-0.02294460	0.049906835	0.012549011

autocorr(model9.4$Sol)
, , (Intercept)

	(Intercept)	gevocab
Lag 0	1.00000000	-0.86375013
Lag 10	-0.01675809	0.01808335
Lag 50	-0.01334607	0.03583885
Lag 100	0.02850369	-0.01102134
Lag 500	0.03392102	-0.04280691

, , gevocab

	(Intercept)	gevocab
Lag 0	-0.863750126	1.0000000000
Lag 10	0.008428317	0.0008246964
Lag 50	0.007928161	-0.0470879801
Lag 100	-0.029552813	0.0237866610
Lag 500	-0.029554289	0.0425010354

The trace plots and histograms, as well as the autocorrelations, indicate that the parameter estimation has converged properly, and that the thinning rate appears to be satisfactory for removing autocorrelation from the estimate values. The model results appear below. First, we should note that the DIC for this random coefficients model is smaller than that of the random intercepts only models above. In addition, the estimate of the random coefficient for vocabulary is 0.2092, with a 95% credible interval of 0.135 to 0.3025. Because this interval does not include 0, we can conclude that the random coefficient is indeed different from 0 in the population, and that the relationship between reading achievement and vocabulary test score varies from one school to another.

```
Iterations = 3001:12991
 Thinning interval  = 10
 Sample size  = 1000

DIC: 42663.14

G-structure:  ~school

        post.mean 1-95% CI u-95% CI eff.samp
school    0.08921   0.0608    0.1256      1000

                ~gevocab

        post.mean 1-95% CI u-95% CI eff.samp
gevocab    0.2092    0.135    0.3025      1000

 R-structure:  ~units

        post.mean 1-95% CI u-95% CI eff.samp
units     3.601     3.508      3.7      1000

 Location effects: geread ~ gevocab

            post.mean 1-95% CI u-95% CI eff.samp  pMCMC
(Intercept)    1.7649    1.4870    1.9891      1000 <0.001 ***
gevocab        0.5501    0.5041    0.5930      1000 <0.001 ***
---
Signif. codes:  0 '***' 0.001 '**' 0.01 '*' 0.05 '.' 0.1 ' ' 1
```

User Defined Priors

Finally, we need to consider the situation in which the user would like to provide her or his own prior distribution information, rather than rely on the defaults established in MCMCglmm. To do so, we will make use of the prior command. In this example, we examine the case where the researcher has informative priors for one of the model parameters. As an example, let us assume that a number of studies in the literature report find a small but consistent positive relationship between reading achievement and a measure of working memory. In order to incorporate this informative prior into a model relating these two variables, while also including the vocabulary score, and accommodating the random coefficient for this variable, we would first need to define our prior, as below. Step one in this process is to create the covariance matrix (var) containing the prior of the fixed effects in the model (intercept and memory). In this case, we set the prior variances of the intercept and the coefficient for memory to 1 and 0.1, respectively. We select a fairly small variance for the working memory coefficient because we have much prior evidence in the literature regarding the anticipated magnitude of this relationship. In addition, we will need to set the priors for the error and random intercept terms. The inverse-Wishart distribution variance structure is used, and here we set the value at 1 with a certainty parameter of nu=0.002.

```
var<-matrix(c(1,0,0,0.1), nrow=2, ncol=2)
prior.model9.4<-list(B=list(mu=c(0,.15), V=var),
G=list(G1=list(V=1, nu=0.002)), R=list(V=1, nu=0.002))
model9.4<-MCMCglmm(geread~npamem, random=~school, data=prime_
time.nomiss, prior=prior.model9.4)
plot(model9.4)
autocorr(model9.4$VCV)
autocorr(model9.4$Sol)
summary(model9.4)
```

The model appears to have converged well, and the autocorrelations suggest that the rate of thinning was appropriate.

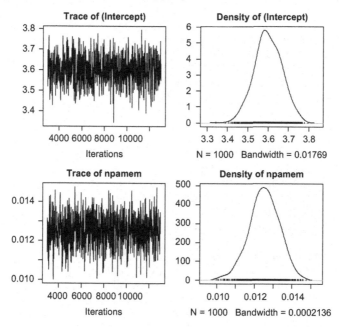

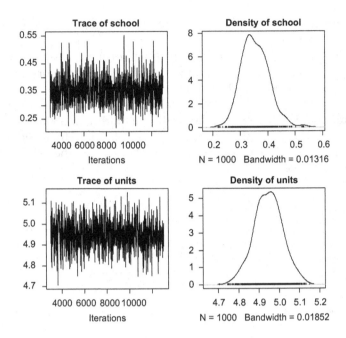

```
autocorr(model9.4$VCV)
, , school

              school            units
Lag 0     1.000000000   -0.019548306
Lag 10    0.046970940   -0.001470008
Lag 50   -0.014670119   -0.051306845
Lag 100   0.020042317    0.013675599
Lag 500  -0.005250327    0.028681171


, , units

              school            units
Lag 0    -0.01954831    1.00000000
Lag 10   -0.01856012    0.03487098
Lag 50    0.05694637    0.01137949
Lag 100  -0.04096406   -0.03780291
Lag 500   0.02678024    0.02372686

autocorr(model9.4$Sol)
, , (Intercept)

            (Intercept)        npamem
Lag 0     1.00000000   -0.640686295
Lag 10    0.02871714    0.004803420
Lag 50   -0.03531602    0.011157302
Lag 100   0.01483541   -0.040542950
```

```
Lag 500  -0.01551577   0.006384573

, , npamem

            (Intercept)          npamem
Lag 0    -0.6406862955  1.000000000
Lag 10   -0.0335729209  -0.022385089
Lag 50    0.0229034652  -0.002681217
Lag 100   0.0007594231   0.008694124
Lag 500   0.0311681203  -0.015291965
```

The summary of the model fit results appear below. Of particular interest is the coefficient for the fixed-effect working memory (npamem). The posterior mean is 0.01266, with a credible interval ranging from 0.01221 to 0.01447, indicating that the relationship between working memory and reading achievement is statistically significant. It is important to note, however, that the estimate of this relationship for the current sample is well below that reported in prior research, and which was incorporated into the prior distribution. In this case, because the sample is so large, the effect of the prior on the posterior distribution is very small. The impact of the prior would be much greater were we working with a smaller sample.

```
Iterations = 3001:12991
Thinning interval  = 10
Sample size  = 1000

DIC: 45908.05

G-structure:  ~school

          post.mean  1-95% CI  u-95% CI  eff.samp
school       0.354     0.2597   0.4508    909.4

R-structure:  ~units

          post.mean  1-95% CI  u-95% CI  eff.samp
units       4.947      4.813    5.096     1000

Location effects: geread ~ npamem

             post.mean  1-95% CI  u-95% CI  eff.samp  pMCMC
(Intercept)   3.59962   3.46707   3.72925     1000   <0.001 ***
npamem        0.01253   0.01094   0.01405     1000   <0.001 ***
---
Signif. codes:  0 '***' 0.001 '**' 0.01 '*' 0.05 '.' 0.1 ' ' 1
```

As a point of comparison, we also fit the model using the default priors in MCMCglmm to see what impact the informative priors had on the posterior distribution. We will focus only on the coefficients for this demonstration, given that they are the focus of the informative priors. For the default priors we obtained the following results.

```
               post.mean 1-95% CI u-95% CI eff.samp  pMCMC
(Intercept)      3.61713  3.47109  3.75194     1000 <0.001 ***
npamem           0.01237  0.01080  0.01417     1000 <0.001 ***
---
Signif. codes:  0 '***' 0.001 '**' 0.01 '*' 0.05 '.' 0.1 ' ' 1
```

Clearly there is virtually no difference in results using the user-supplied informative and the default noninformative priors, demonstrating that when the sample size is large the selection of priors will often not have much bearing on the final results of the analysis.

MCMCglmm for a Dichotomous Dependent Variable

The MCMCglmm library can also be used to fit multilevel models in which the outcome variable is dichotomous in nature. In most respects, the use of the functions from this library will be very similar to what we have seen with a continuous outcome, as described previously. Therefore, we will focus on aspects of model fitting that differ from what we have seen up to this point. Our first example involves fitting a model for a dichotomous dependent variable using Bayesian multilevel logistic regression. Specifically, the model of interest involves predicting whether or not a student receives a passing score on a state math assessment (score2) as a function of their number sense (numsense) score on a formative math assessment. Following is the R code for fitting this model, and then requesting the plots and output.

```
mathfinal.nomiss<-na.omit(mathfinal)
model9.5<-MCMCglmm(score2~numsense, random=~school,
family="ordinal", data=mathfinal,)
plot(model9.5)
autocorr(model9.5$VCV)
autocorr(model9.5$Sol)
summary(model9.5)
```

The default prior parameters are used, and the family is defined as ordinal. In other respects, the function call is identical to that for the continuous outcome variables that were the focus of the earlier part of this chapter. The output from R appears below. From the trace plots and histograms, we can see that convergence was achieved for each of the model parameters, and the autocorrelations show that our rate of thinning is sufficient.

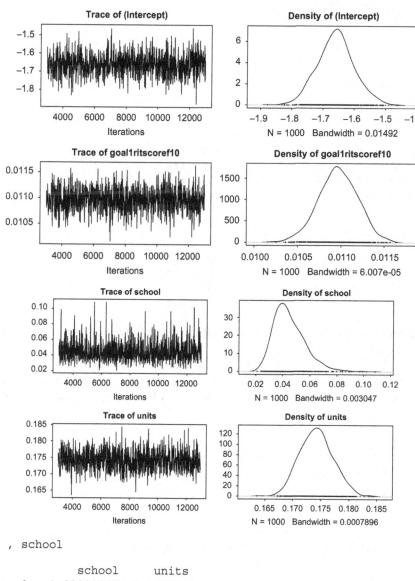

```
,  ,  school

               school          units
Lag 0     1.000000000    0.24070410
Lag 10    0.016565749    0.02285168
Lag 50    0.012622856    0.02073446
Lag 100   0.007855806    0.02231629
Lag 500   0.007233911    0.01822021
```

```
, , units

          school      units
Lag 0    0.24070410 1.00000000
Lag 10   0.02374442 0.00979023
Lag 50   0.02015865 0.00917857
Lag 100  0.01965188 0.00849276
Lag 500  0.01361470 0.00459030


, , (Intercept)

          (Intercept) numsense
Lag 0     1.00000000      -0.09818969
Lag 10    0.00862290      -0.00878574
Lag 50    0.00688767      -0.00707115
Lag 100   0.00580816      -0.00603118
Lag 500   0.00300539      -0.00314349

, , numsense

          (Intercept) numsense
Lag 0    -0.09818969    1.0000000
Lag 10   -0.00876214    0.00894084
Lag 50   -0.00704441    0.00723130
Lag 100  -0.00594502    0.00618679
Lag 500  -0.00315547    0.00328528
```

In terms of model parameter estimation results, the number sense score was found to be statistically significantly related to whether or not a student received a passing score on the state mathematics assessment. The posterior mean for the coefficient is 0.04544, indicating that the higher an individual's number sense score, the greater the likelihood that (s)he will pass the state assessment.

```
Iterations = 3001:12991
 Thinning interval  = 10
 Sample size  = 1000

 DIC: 6333.728

 G-structure:  ~school

        post.mean 1-95% CI u-95% CI eff.samp
school    0.3586    0.1534    0.5874       127

 R-structure:  ~units

        post.mean 1-95% CI u-95% CI eff.samp
units     1.769     1.182     2.603     8.045

 Location effects: score2 ~ numsense
```

```
            post.mean   1-95% CI   u-95% CI  eff.samp   pMCMC
(Intercept) -12.42531  -14.23106  -10.67995     10.04  <0.001  ***
numsense      0.06306    0.05471    0.07250     11.09  <0.001  ***
---
Signif. codes:  0 '***' 0.001 '**' 0.01 '*' 0.05 '.' 0.1 ' ' 1
```

Exactly the same command sequence that was used here would also be used to fit a model for an ordinal variable with more than two categories.

MCMCglmm for a Count-Dependent Variable

As may be obvious at this point, using the R function MCMCglmm, it is possible to fit the same types of multilevel models in the Bayesian context that we were able to fit using REML with lme and lmer, including for count data. As we saw in Chapters 7 and 8, Poisson regression is the methodology that is typically used with such data. In order to demonstrate the modeling of a count outcome in the Bayesian context, we will revisit the example that was our focus at the end of Chapter 8. Recall that in this example the dependent variable was the number of cardiac warning incidents, such as chest pain, shortness of breath, and dizzy spells, that occurred over a six-month period for each of 1,000 patients who were being treated in 110 cardiac rehabilitation facilities. Study participants were randomly assigned to either a new exercise treatment program, or to the standard treatment. At the end of the study, the researchers were interested in comparing the frequency of cardiac warning signs between the two treatments, while controlling for the sex of the patient. Given that the frequency of the cardiac warning signs was very small across the six-month period, Poisson regression was deemed to be the optimal analysis for answering the research question regarding whether the new treatment resulted in better outcomes than the old. In order to fit such a model using MCMCglmm, we use the following commands.

```
attach(heartdata)
model9.6<-MCMCglmm(heart~trt+sex, random=~rehab,
family="poisson", data=heartdata)
plot(model9.6)
autocorr(model9.6$VCV)
autocorr(model9.6$Sol)
summary(model9.6)
```

The key subcommand here is family="poisson", which indicates that Poisson regression is to be used. In all other respects, the syntax is identical to that used for the continuous and dichotomous variable models. The trace plots and histograms for assessing model convergence appear below.

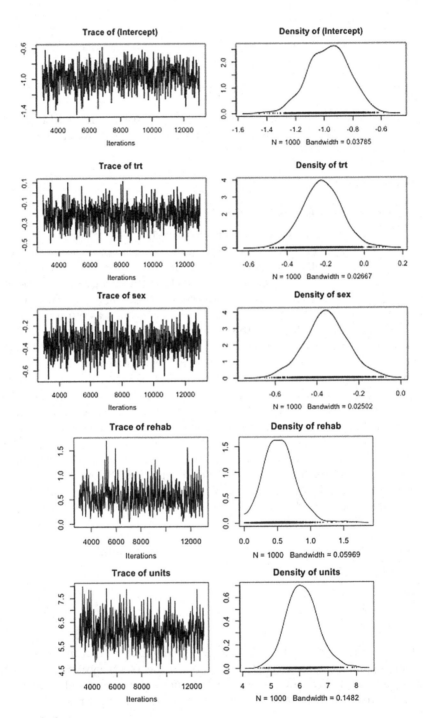

, , rehab

rehab   units

```
Lag 0      1.000000000  -0.0117468496
Lag 10     0.004869176  -0.0067184848
Lag 50     0.000957586   0.0009480950
Lag 100    0.009502289   0.0004500062
Lag 500   -0.009067234   0.0028298115

, , units

                    rehab          units
Lag 0      -0.00117468  1.00000000000
Lag 10     -0.00076201  0.00425938977
Lag 50      0.00013997  0.00065406398
Lag 100     0.00035229  0.00079448090
Lag 500     0.00024450  0.00011262469

, , (Intercept)

            (Intercept)          trt             sex
Lag 0      1.00000000000   0.0002158950   0.00171708145
Lag 10     0.00268697330   0.0003701961   0.00100571606
Lag 50     0.00058216804  -0.0001337596   0.00030117833
Lag 100    0.00009689295   0.0003694162  -0.00033360474
Lag 500    0.00002480209   0.0003205542  -0.00003672349

, , trt

            (Intercept)          trt             sex
Lag 0       0.0002158950  1.0000000000   0.0007192931
Lag 10      0.0010499669  0.0005487463  -0.0001185169
Lag 50     -0.0001931866 -0.0002920215  -0.0004492621
Lag 100    -0.0002697260 -0.0001977527  -0.0001267768
Lag 500     0.0002656560 -0.0002109309  -0.0005854029

, , sex

            (Intercept)          trt             sex
Lag 0       0.00171708145  0.0007192931  1.00000000000
Lag 10      0.00037221141  0.0004940633  0.00058844721
Lag 50     -0.00064352200  0.0002252359  0.00006823018
Lag 100     0.00009610112  0.0008764231 -0.00042699447
Lag 500    -0.00016594722 -0.0001365390  0.00010049097
```

An examination of the trace plots and histograms shows that the parameter estimation converged appropriately. In addition, the autocorrelations are sufficiently small for each of the parameters so that we can have confidence in our rate of thinning. Therefore, we can move to the discussion of the model parameter estimates, which appear below.

```
Iterations = 3001:12991
 Thinning interval  = 10
 Sample size  = 1000
```

```
DIC: 2735.293

G-structure:  ~rehab

       post.mean 1-95% CI u-95% CI eff.samp
rehab     0.5414    0.1022    1.009     1000

R-structure:  ~units

       post.mean 1-95% CI u-95% CI eff.samp
units     6.102     5.074     7.324     1000

Location effects: heart ~ trt + sex

              post.mean 1-95% CI u-95% CI eff.samp  pMCMC
(Intercept)   -0.96877 -1.23267 -0.68596    1000 <0.001 ***
trt           -0.21909 -0.40769 -0.01448    1000   0.03 *
sex           -0.35585 -0.57662 -0.16348    1000 <0.001 ***
---
Signif. codes:  0 '***' 0.001 '**' 0.01 '*' 0.05 '.' 0.1 ' ' 1
```

In terms of the primary research question, the results indicate that the frequency of cardiac risk signs was lower among those in the treatment condition than those in the control, when accounting for the participants' sex. In addition, there was a statistically significant difference in the rate of risk symptoms between males and females. With respect to the random effects, the variance in the outcome variable due to rehabilitation facility, as well as the residual, were both significantly different from 0. The posterior mean effect of the rehab facility was 0.5414, with a 95% credibility interval of 0.1022 to 1.009. This result indicates that symptom frequency does differ among the facilities.

We may also be interested in examining a somewhat more complex explanation of the impact of treatment on the rate of cardiac symptoms. For instance, there is evidence from previous research that the number of hours the facilities are open may impact the frequency of cardiac symptoms, by providing more, or less, opportunity for patients to make use of their services. In turn, if more participation in rehabilitation activities is associated with the frequency of cardiac risk symptoms, we might expect the hours of operation to impact them. In addition, it is believed that the impact of the treatment on the outcome might vary among rehabilitation centers, leading to a random coefficients model. The R commands to fit the random coefficients (for treatment) model, with a level-2 covariate (hours of operation) appear below, followed by the resulting output. Note that, as we have seen in previous examples in this chapter, in order to specify a random coefficients model, we include the variables of interest (rehab and hours) in the random statement.

```
model9.7<-MCMCglmm(heart~trt+sex+hours, random=~rehab+trt,
family="poisson", data=heartdata)
```

```
plot(model9.7)
autocorr(model9.7$VCV)
autocorr(model9.7$Sol)
summary(model9.7)
```

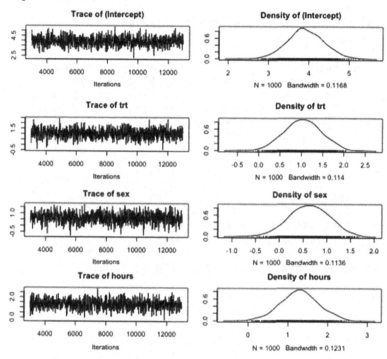

, , rehab

```
            rehab          trt          units
Lag 0    1.00000000 -0.266851868 -0.16465715
Lag 10   0.00378468 -0.021179331 -0.01630321
Lag 50   0.00190117 -0.018558364 -0.01613084
Lag 100  0.00215891 -0.015675323 -0.02000173
Lag 500  0.00134143 -0.004070154 -0.02503848
```

, , trt

```
            rehab          trt          units
Lag 0    -0.2668519 1.000000000 -0.11400674
Lag 10   -0.02075032 0.00740803 -0.01180379
Lag 50   -0.02016743 0.00702657 -0.01022964
Lag 100  -0.02234751 0.00681188 -0.00740961
Lag 500  -0.02362045 0.00552936  0.00579908
```

, , units

	rehab	trt	units
Lag 0	-0.16465715	-0.11400674	1.00000000
Lag 10	-0.01475261	-0.01183465	0.00967745
Lag 50	-0.01055799	-0.01229537	0.00862259
Lag 100	-0.01106924	-0.01174882	0.00749455
Lag 500	-0.00619535	-0.00694897	0.00282392

, , (Intercept)

	(Intercept)	trt	sex	hours
Lag 0	1.00000000	0.10756409	0.1718756	-0.06742879
Lag 10	0.00337255	0.01117497	0.0181841	0.01915046
Lag 50	0.00369810	0.00898286	0.0211193	0.03380228
Lag 100	0.00373422	0.01070748	0.0187302	0.01015611
Lag 500	0.00290292	0.00500545	0.0202833	0.04459613

, , trt

	(Intercept)	trt	sex	hours
Lag 0	0.10756409	1.00000000	0.1465124	0.09764357
Lag 10	0.01056947	0.00361113	0.0136449	0.01181678
Lag 50	0.00760802	0.00350369	0.0153875	0.00992076
Lag 100	0.00924803	0.00346671	0.0141195	0.00557474
Lag 500	0.00528078	0.00177952	0.0038593	0.00150113

, , sex

	(Intercept)	trt	sex	hours
Lag 0	0.1718756	0.14651224	1.00000000	0.15838758
Lag 10	0.0191615	0.01300171	0.00316606	0.01377718
Lag 50	0.0208031	0.01265926	0.00248169	0.00786715
Lag 100	0.0189574	0.00596110	0.00287549	0.01078291
Lag 500	0.0115626	0.00662074	0.00111150	0.00931525

, , hours

	(Intercept)	trt	sex	hours
Lag 0	-0.06742879	0.09764357	0.15838758	1.00000000
Lag 10	0.00447807	0.00877490	0.01110063	0.01592455
Lag 50	0.00662157	0.01028671	0.00917307	0.01552734
Lag 100	0.00511239	0.00835433	0.00759920	0.00231364
Lag 500	0.00526215	0.01093175	0.00397990	0.00702958

The trace plots and histograms reveal that estimation converged for each of the parameters estimated in the analysis, and the autocorrelations of estimates are small. Thus, we can move on to interpretation of the parameter estimates. The results of the model fitting revealed several interesting patterns. First, the random coefficient term for treatment was statistically significant, given that the credible interval ranged between 5.421 and 7.607, and did not include 0. Thus, we can conclude that the impact of treatment on the number of cardiac symptoms differs from one rehabilitation center to the next. In addition, the variance in the outcome due to rehabilitation center was also different from 0, given that the confidence interval for rehab was 0.1953 to 1.06. Finally, treatment and sex were negatively statistically significantly related to the number of cardiac symptoms, as they were for Model9.7, and the centers' hours of operation were not related to the number of cardiac symptoms.

```
Iterations = 3001:12991
Thinning interval  = 10
Sample size  = 1000

DIC: 2731.221

G-structure:   ~rehab

       post.mean 1-95% CI u-95% CI eff.samp
rehab    0.5708    0.1291    1.017    286.6

            ~trt

      post.mean 1-95% CI u-95% CI eff.samp
trt    0.5944 0.000127    4.923    5.552

  R-structure:   ~units

       post.mean 1-95% CI u-95% CI eff.samp
units     5.555     0.55    7.332    8.122

Location effects: heart ~ trt + sex + hours

              post.mean   1-95% CI   u-95% CI eff.samp   pMCMC
(Intercept) -0.976812 -1.290830 -0.701594    229.5 <0.001 ***
trt         -0.222578 -0.413242 -0.019580    440.0  0.024 *
sex         -0.365893 -0.556338 -0.167009    617.9 <0.001 ***
hours        0.008937 -0.236606  0.272954   1000.0  0.936
---
Signif. codes:  0 '***' 0.001 '**' 0.01 '*' 0.05 '.' 0.1 ' ' 1
```

Summary

The material presented in Chapter 9 represents a marked departure from that presented in the first eight chapters of the book. In particular, methods presented in the earlier chapters were built upon a foundation of maximum likelihood estimation. Bayesian modeling, which is the focus of Chapter 9, leaves likelihood-based analyses behind, and instead relies on MCMC to derive a posterior distribution for model parameters. More fundamentally, however, Bayesian statistics is radically different from likelihood-based frequentist statistics in terms of the way in which population parameters are estimated. In the latter they take single values, whereas Bayesian estimation estimates population parameters as distributions, so that the sample-based estimate of the parameter is the posterior distribution obtained using MCMC, and not a single value calculated from the sample.

Beyond the more theoretical differences between the methods described in Chapter 9 and those presented earlier, there are also very real differences in terms of application. Analysts using Bayesian statistics are presented with what is, in many respects, a more flexible modeling paradigm that does not rest on the standard assumptions that must be met for the successful use of maximum likelihood, such as normality. At the same time, this greater flexibility comes at the cost of greater complexity in estimating the model. From a very practical viewpoint, consider how much more is involved when conducting the analyses featured in Chapter 9 as compared to those featured in Chapter 3. In addition, interpretation of the results from Bayesian modeling requires more from the data analyst, in the form of ensuring model convergence, deciding on the length of the chains, and the degree of thinning that should be done, as well as what summary statistic of the posterior will be used to provide a single-parameter estimate. Last, and not at all least, the analyst must consider what the prior distributions of the model parameters should be, knowing that particularly for small samples, the choice will have a direct bearing on the final parameter estimates that are obtained.

In spite of the many complexities presented by Bayesian modeling, it is also true that Bayesian models offer the careful and informed researcher a very flexible and powerful set of tools. In particular, as we discussed at the beginning of the chapter, Bayesian analysis, including multilevel models, allow for greater flexibility in model form, do not require distributional assumptions, and may be particularly useful for smaller samples. Therefore, we can recommend this approach wholeheartedly, with the realization that the interested researcher will need to invest greater time and energy into deciding on priors, and determining when/if a model has converged and what summary statistic of the posterior distribution is most appropriate. Given this investment, you have the potential for some very useful and flexible models that may work in situations where standard likelihood-based approaches do not.

10

Advanced Issues in Multilevel Modeling

The purpose of this chapter is to introduce a wide array of topics in the area of multilevel analysis that do not fit neatly in any of the other chapters in the book. We refer to these as advanced issues because they represent extensions, of one kind or another, on the standard multilevel modeling framework that we have discussed heretofore. In this chapter we will describe how the estimation of multilevel model parameters can be adjusted for the presence of outliers using robust or rank-based methods. As we will see, such approaches provide the researcher with powerful tools for handling situations when data do not conform to the distributional assumptions underlying the models that we discussed in Chapters 3 and 4. We will then turn our attention to the problem of fitting multilevel models when there are a large number of predictor variables. Such data structure can make it difficult to obtain reasonable parameter and standard error estimates. In order to address this issue, penalized estimation procedures can be used in order to identify only those predictors that are the most salient for a given model and set of data. We will then turn our attention to multivariate multilevel data problems, in which there are multiple dependent variables. This situation corresponds to multivariate analysis of variance and multivariate regression in the single-level modeling framework. Next, we will examine multilevel generalized additive models, which can be used in situations where the relationship between a predictor and an outcome variable is nonlinear. We will finish out the chapter with a discussion of predicting level-2 outcome variables with level-1 independent variables, and with a description of approaches for power analysis/sample size determination in the context of multilevel modeling.

Robust Statistics in the Multilevel Context

The focus of much of this book has been on situations in which the data to be analyzed can be said to be well behaved, in the sense that they conform to some known distribution. For example, in Chapters 2 through 5, the data are assumed to be normally distributed. In Chapter 1, we discussed methods for assessing this assumption using the residuals associated with regression modeling. In Chapters 7 and 8, our focus turned to data that have

known non-normal distributions (e.g. dependent variables that are dichotomous, count data with mean equal to the variance, overdispersed count data,
etc.). However, in many real-world applications, the data do not conform
to these distributional assumptions. For example, when outliers are present, data may be skewed, rather than normally distributed. In addition, the
presence of outliers can impact the accuracy of parameter estimates, such
as coefficients, as well as their standard errors (Finch, 2017; Kloke, McKean,
and Rashid, 2009; Pinheiro, Liu, and Wu, 2001). In order to deal with the
problem of outliers, alternative methods have been suggested. In particular,
we will examine a set of methods that assume the data come from heavy-
tailed distributions, which might arise when outliers are present, as well as
a technique based on ranks of the original scores, upon which outliers have
a relatively lighter impact.

Identifying Potential Outliers in Single-Level Data

The presence of outliers can have a deleterious impact on the estimation of
model parameters and standard errors for multilevel models, just as it can for
single-level models (Staudenmayer, Lake, and Wand, 2009). Thus, researchers and data analysts should screen their data for the presence of outliers,
prior to conducting their target analyses. If outliers are detected, a decision
must be made regarding how to deal with them. The general state of the
art at this time is to attempt modeling of the data as it is, outliers included,
rather than removing outlying observations and fitting a standard multilevel model on the truncated dataset (Staudenmayer et al., 2009). Given this
emphasis on using an appropriate model that accommodates outliers, the
researcher must have some sense as to whether outliers are in fact present, so
that she can select the optimal analytic approach. In this and the next section
of the chapter we describe some of the more common approaches for identifying outliers, first for single-level data, and then in the context of multilevel
data. We will then focus on multilevel modeling strategies that are available
to the researcher, given that outliers are present in the data.

Several of the more common outlier detection methods that are used
with multilevel data have direct antecedents in the single-level regression
literature. One of these, Cook's distance (D), compares residuals (difference
between observed and model-predicted outcome variables) for individual
cases when another observation is included in the data, versus when it is
removed, while correcting for the leverage of the data point. Leverage is a
measure of how far an individual's set of values on the independent variables is from the mean of these values. When there are multiple independent
variables, the leverage statistic measures the distance between an individual's values on the independent variables and the set of means for those

independent variables, which is commonly referred to as the centroid. Large leverage values indicate that an observation may have a great impact on the predicted values produced by the model. Another important measure of an observation's impact on a regression model is Cook's D, which is a general-use statistic designed to identify data points that have an outsized impact on the fit of the model. For standard regression models, D is calculated as

$$D_i = \frac{\sum_{i=1}^{N}(e_i - e_{ij})^2}{kMSr} \tag{10.1}$$

where
$\quad e_i \quad$ = residual for observation i for model containing all observations
$\quad e_{ij} \quad$ = residual for observation i for model with observation j removed
$\quad k \quad$ = number of independent variables
$\quad MSr$ = mean square of the residuals

There are no hard and fast rules for how large D_i should be in order for us to conclude that it represents an outlying observation. Fox (2016) recommends that the data analyst flag observations that have D_i values that are unusual when compared to the rest of those in the dataset, and this is the approach that we will recommend as well.

Another influence diagnostic, which is closely related to D_i, is DFFITS. This statistic compares the predicted value for individual i when the full dataset is used ($\hat{y}_i$), against the prediction for individual i when individual j is dropped from the data ($\hat{y}_{ij}$). For individual j, DFFITS is calculated as:

$$DFFITS_j = \frac{\hat{y}_i - \hat{y}_{ij}}{\sqrt{MSE_j h_j}} \tag{10.2}$$

where
$\quad MSE_j \quad$ = mean squared error when observation j is dropped from the data
$\quad h_j \quad$ = leverage value for observation j

As was the case with D_i, there are no hard and fast rules about how large $DFFITS_j$ should be in order to flag observation j as an outlier. Rather, we examine the set of $DFFITS_j$ and focus on those that are unusually large (in absolute value) when compared to the others. One final outlier detection tool for single-level models that we will discuss here is the *COVRATIO*, which measures the impact of an observation on the precision of model estimates. For individual i, this statistic is calculated as

$$COVRATIO_i = \frac{1}{(1-h_i)}\left(\frac{n-k-2+E_i^{*2}}{n-k-1}\right)^{k+1} \tag{10.3}$$

where
E_i^* = studentized residual
n = total sample size

Fox (2016) states that *COVRATIO* values greater than 1 improve the precision of the model estimates, whereas those with values lower than 1 decrease the precision of the estimate. Clearly, it is preferable for observations to increase model precision, rather than decrease it.

Identifying Potential Outliers in Multilevel Data

Detection of potential outliers in the context of multilevel data is, in most respects, similar to the single-level methods described above. The primary difference with multilevel data is in how observations are defined when it comes to removing them in the calculations. In the single-level case, a single observation, corresponding to one member of the sample, is removed for DFFITS, as an example. However, for multilevel data the observation in question is typically associated with level 2, rather than level 1. Thus, when calculating statistics such as D_i, the data to be removed corresponds to an entire cluster, and not just to a single data point. Furthermore, for multilevel models, D_i provides information about outliers with respect to the fixed-effects portion of the model only.

As was the case for single-level models, D_i for multilevel models is based on the leverage values for the observations. Based on work by Demidenko and Stukel (2005), the fixed-effects leverage value for individual i is calculated as:

$$H_{i,\text{Fixed}} = X_i \left(\sum_{i=1}^{N} X_i' V_i^{-1} X_i \right)^{-1} X_i' V_i^{-1} \qquad (10.4)$$

where
X_i = matrix of fixed effects for subject i
V_i = covariance matrix of the fixed effects for subject i
N = number of level-2 units

Note that in Equation (10.4), the subject refers to the level-2 grouping variable. Similarly, the leverage values based on the random effects for subjects in the sample can be expressed as

$$H_{i,\text{Random}} = Z_i D Z_i' V_i^{-1} \left(I - H_{i,\text{Fixed}} \right) \qquad (10.5)$$

where
Z_i = matrix of random effects for subject i
D = covariance matrix of the random effects for subject i.

The actual fixed- and random-effects leverage values correspond to the diagonal elements of $H_{i,\text{Fixed}}$ and $H_{i,\text{Random}}$, respectively. The multilevel analog of Cook's D_i can then be calculated using the following equation:

$$D_{Mi} = \frac{1}{mS_e^2} r_i' \left(I - H_{i,\text{Fixed}}\right)^{-1} V_i^{-1} X_i \left(\sum_{i=1}^{N} X_i' V_i^{-1} X_i\right)^{-1} X_i' V_i^{-1} \left(I - H_{i,\text{Fixed}}\right)^{-1} r_i \quad (10.6)$$

where
    m = number of fixed-effects parameters
    S_e^2 = estimated error variance
    I = identity matrix

$$r_i = y_i - \hat{y}_i$$

    y_i = observed value of dependent variable for observation i
    $\hat{y}_i$ = model-predicted value of dependent variable for observation i

The interpretation of D_{Mi} is similar to that for D_i, in that individual values departing from the main body of values in the sample are seen as indicative of potential outliers.

In addition to the influence statistics at each level, and D_{Mi}, there is also a multilevel analog for DFFITS, known as MDFFITS. As in the single-level case, MDFFITS is a measure of the amount of change in the model-predicted values for the sample when an individual is removed versus when they are retained. This measure is considered an indicator of potential outliers with respect to the fixed effects, given the random-effect structure present in the data. Likewise, the COVRATIO statistic can also be used to identify potential outliers with respect to their impact on the precision with which model parameters are estimated. The same rules of thumb for interpreting MDFFITS and the COVRATIO that were described for single-level models also apply in the context of multilevel modeling.

With respect to the random effects, the relative variance change (RVC; Dillane, 2005) statistic can be used to identify potential outliers. The RVC for one of the random variance components for a subject in the sample is calculated as

$$\text{RVC}_i = \frac{\hat{\theta}_i}{\hat{\theta}} - 1 \qquad (10.7)$$

where
    $\hat{\theta}$ = variance component of interest (e.g. residual, random intercept) estimated using full sample
    $\hat{\theta}_i$ = variance component of interest estimated excluding subject i

When RVC is close to 0, the observation does not have much influence on the variance component estimate; i.e. is not likely to be an outlier. As with the other statistics for identifying potential outliers, there is not a single agreed-upon cut-value for identifying outliers using RVC. Rather, observations that have unusual such values when compared to the full sample warrant special attention in this regard.

Identifying Potential Multilevel Outliers Using R

In order to demonstrate the functions in R that can be used to identify potential outliers, we will use the dental data that is part of the heavy library. As described in the documentation for this package, the data were originally reported in Potthoff and Roy (1964), and consist of dental measurements made on 11 females and 16 males at ages 8, 10, 12, and 14. The outcome of interest was the distance, measured in millimeters, from the center of the pituitary to the pterygomaxillary fissure. The two independent variables of interest were subject gender (females = 1, males = 0) and age. In order to gain a better understanding of the data, we might first examine a simple histogram.

```
qplot(dental$distance, geom="histogram", main="Histogram of
Distance", xlab="Distance") + theme(plot.title = element_
text(hjust = 0.5))
```

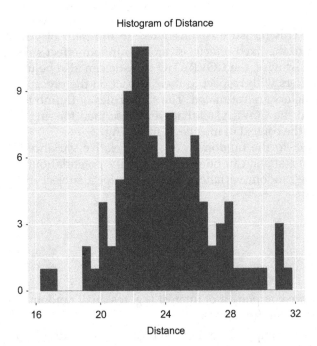

We can see that there are some individual measurements at both ends of the distance scale that are separated from the main body of measurements. It's important to remember that this graph does not organize the observations by the individual on whom the measurements were made, whereas in the context of multilevel data we are primarily interested in examining the data at that level. Nonetheless, this type of exploration does provide us with some initial insights into what we might expect to find with respect to outliers moving forward.

Next, we can examine a boxplot of the measurements by subject gender.

```
ggplot(dental, aes(x=Sex, y=distance)) + geom_boxplot()+
labs(title="Boxplot of Distance by Sex",x ="Distance", y =
"Sex")+ theme(plot.title = element_text(hjust = 0.5))
```

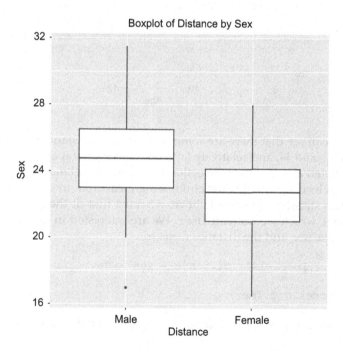

The distribution of distance is fairly similar for males and females, though there is one measurement for a male subject that is quite small when compared with the other male distances, and indeed is small when compared to the female measurements as well. For this sample, typical male distance measurements are somewhat larger than are those for females. Finally, we can examine the relationship between age and distance using a scatterplot.

```
ggplot(dental, aes(x=age, y=distance)) + geom_point()+
labs(title="Scatterplot of Distance by Age",x ="Age", y =
"Distance")+ theme(plot.title = element_text(hjust = 0.5))
```

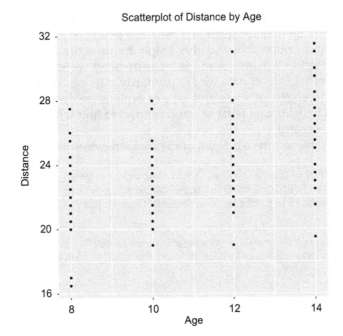

Scatterplot of Distance by Age

Here we can see that there are some relatively low-distance measurements at ages 8, 12, and 14, and relatively large measurements at ages 10 and 12.

As we noted above, outlier detection in the multilevel modeling context is focused on level 2, which is the child being measured. In order to calculate the various statistics described above, we must first fit the standard multi-level model, which we do with lmer. We are interested in the relationships between age, sex, and the distance measure.

```
model10.1.lmer<-lmer(distance ~ age + Sex + (1 | Subject),
dental)
summary(model10.1.lmer)
```

Below is the portion of the output showing the random- and fixed-effects estimates.

```
Random effects:
Groups     Name            Variance  Std.Dev.
Subject    (Intercept)     3.267     1.807
Residual                   2.049     1.432
Number of obs: 108, groups:  Subject, 27

Fixed effects:
             Estimate Std. Error        df t value Pr(>|t|)
(Intercept)  17.70671    0.83392  99.35237  21.233  < 2e-16 ***
```

```
age              0.66019     0.06161 80.00000  10.716  < 2e-16 ***
SexFemale       -2.32102     0.76142 25.00000  -3.048  0.00538 **
---
Signif. codes:  0 '***' 0.001 '**' 0.01 '*' 0.05 '.' 0.1 ' ' 1
```

These results indicate that measurements made on the children when they were older were larger, and measurements made on females were smaller.

In order to obtain the influence statistics for this problem, we will use the HLMdiag package in R. The following two commands are used to calculate the outlier diagnostics that are described above. The first line involves the deletion of each subject in the sample, in turn, and then the refitting of model10.1.lmer. The second line then commands R to calculate the influence diagnostics, and save them in the output object subject. diag.1.

```
library(HLMdiag)

subject.del.1 <- case_delete(model = model10.1.lmer, group =
"Subject")
subject.diag.1 <- diagnostics(subject.del.1)
```

We can examine the individual statistics within this object.

```
   subject.diag.1
$`fixef_diag`
     IDS      COOKSD     MDFFITS    COVTRACE   COVRATIO
1    M01  0.059430549 0.058372709 0.001283859 0.9992782
2    M02  0.017644430 0.016373372 0.153849114 1.1605606
3    M03  0.004288299 0.003930055 0.191718307 1.2028549
4    M04  0.047790333 0.045490961 0.122488826 1.1263693
5    M05  0.028992132 0.027350696 0.105918181 1.1084350
6    M06  0.012432106 0.011444600 0.183111466 1.1930342
7    M07  0.011856240 0.010908774 0.183384134 1.1935544
8    M08  0.018070149 0.017113021 0.152846918 1.1591818
9    M09  0.013029622 0.016481864 0.059973890 0.9113512
10   M10  0.129831010 0.151859927 0.276663246 0.7305381
11   M11  0.025922322 0.024178483 0.168673525 1.1770987
12   M12  0.018243029 0.017244419 0.188554582 1.1992358
13   M13  0.219383621 0.271711664 0.055608520 0.9199265
14   M14  0.002429481 0.002285124 0.214674902 1.2290509
15   M15  0.033221623 0.031746926 0.169870517 1.1781381
16   M16  0.025888407 0.024358538 0.126292525 1.1299977
17   F01  0.026282969 0.023991066 0.201711812 1.2127109
18   F02  0.003743791 0.003422279 0.255982553 1.2753945
19   F03  0.016469517 0.014842882 0.218707274 1.2321454
20   F04  0.052586476 0.048732087 0.134238794 1.1371272
```

```
21 F05 0.019281430 0.018289702 0.241688289 1.2582548
22 F06 0.033063598 0.030097037 0.193783496 1.2038339
23 F07 0.002781395 0.002520583 0.258639982 1.2785572
24 F08 0.035716391 0.033738879 0.220283123 1.2334511
25 F09 0.041773274 0.038463687 0.182647207 1.1912499
26 F10 0.172655722 0.187556117 0.187313026 0.8116673
27 F11 0.134819209 0.140361133 0.097798809 0.8961567


$varcomp_diag
        IDS        sigma2              D11
M01 M01 -0.001876300 -0.056743115
M02 M02  0.023443524  0.009967692
M03 M03  0.012552130  0.039204688
M04 M04 -0.002001208  0.011194962
M05 M05  0.008555067 -0.005872565
M06 M06  0.038456882  0.015265800
M07 M07  0.027199899  0.023727561
M08 M08 -0.008971702  0.033333394
M09 M09 -0.241337636  0.085721068
M10 M10  0.029978359 -0.235840135
M11 M11  0.020531028  0.020415631
M12 M12  0.009505588  0.039682505
M13 M13 -0.225614426  0.076558841
M14 M14  0.022773285  0.044510352
M15 M15 -0.001355337  0.037240673
M16 M16  0.028174805 -0.008949734
F01 F01  0.016375549  0.022906873
F02 F02  0.031952503  0.041449235
F03 F03  0.022812888  0.027577068
F04 F04  0.033338696 -0.026626936
F05 F05  0.015156927  0.045817231
F06 F06  0.025880757  0.011610962
F07 F07  0.034511615  0.041047801
F08 F08  0.007504013  0.039603874
F09 F09  0.015156926  0.013292892
F10 F10  0.023958384 -0.196924903
F11 F11  0.028951674 -0.151253263
```

However, given that identification of potential outliers is done by finding observations with values that are different from those of the bulk of the sample, it may be easier to do this graphically. Thus, we will graph the multilevel Cook's D and MDFFITS using the dotplot _ diag function in HLMdiag.

```
dotplot_diag(x = COOKSD, index = IDS, data = subject.
diag[["fixef_diag"]],name = "cooks.distance", modify =
FALSE,xlab = "Subject", ylab = "Cook's Distance")
```

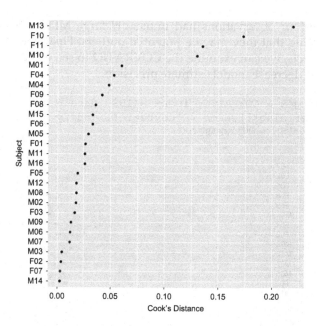

Based on this graph, it appears that subjects M13, F10, F11, and M10 all have Cook's D values that are unusual when compared to the rest of the sample.

```
dotplot_diag(x = MDFFITS, index = IDS, data = subject.
diag[["fixef_diag"]],name = "mdffits", modify = FALSE,xlab =
"Subject", ylab = "MDFFITS")
```

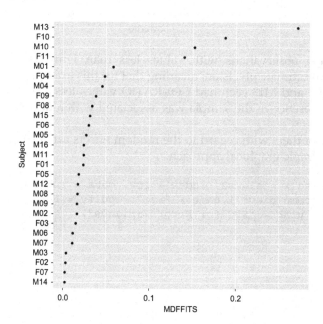

The same four observations had unusually large MDFFITS values, further suggesting that they may be potential outliers. Finally, with regard to the fixed effects, we can examine the COVRATIO to see whether any of the observations decrease model estimate precision.

```
dotplot_diag(x = COVRATIO, index = IDS, data = subject.
diag[["fixef_diag"]],name = "covratio", modify = FALSE,xlab =
"Subject", ylab = "COVRATIO")
```

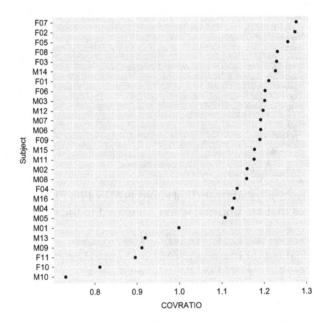

Recall that observations with values less than 1 are associated with decreases in estimate precision, meaning that for this example subjects M10, F10, F11, M9, and M13 each had COVRATIO values less than 1, indicating that their presence in the sample was associated with a decrease in model precision.

Potential outliers with regard to the random effects can also be identified using this simple graphical approach.

```
dotplot_diag(x = sigma2, index = IDS, data = subject.
diag[["varcomp_diag"]],name = "rvc", modify = FALSE,xlab =
"Subject", ylab = "RVC for Error Variance")
```

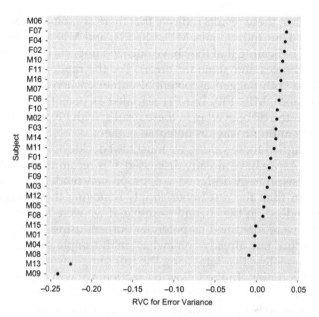

As with Cook's D and MDFFITS, observations with unusual RVC values are identified as potential outliers, meaning that M9 and M13 would be likely candidates with regard to the error variance. In terms of the random intercept variance component, M10, F10, and F11 were the potential outlying observations.

```
dotplot_diag(x = D11, index = IDS, data = subject.
diag[["varcomp_diag"]],name = "rvc", modify = FALSE,xlab =
"Subject", ylab = "RVC for Intercept Variance")
```

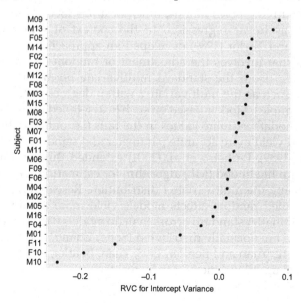

Taken together, these results indicate that there are some potential outliers in the sample. Four individuals, M13, F10, F11, and M10 all were associated with statistics indicating that they have outsized impacts on the fixed-effects parameter estimates and standard errors. For the random error effect, M9 and M13 were both flagged as potential outliers, whereas for the random intercept variance, M10, F10, and F11 signaled as potential outliers. Notice that there is quite a bit of overlap in these results, with F10, F11, M10, and M13 showing up in multiple ways as potential outliers.

Having identified these values, we must next decide how we would like to handle them. As we have already discussed, it is recommended that, if possible, an analysis strategy designed to appropriately model data with outliers be used, rather than the researcher simply removing them (Staudenmayer et al., 2009). Indeed, in the current example, if we were to remove the four subjects who were identified as potential outliers using multiple statistics, we would reduce our already somewhat small sample size from 27 down to 23. Thus, modeling the data using the full sample would seem to be the more attractive option. In the following section, we will discuss estimators that can be used for modeling multilevel data when outliers may be present. After discussing these from a theoretical perspective, we will then demonstrate how to employ them using R.

Robust and Rank-Based Estimation for Multilevel Models

Although there has been much work done in the development of models to deal with outlying observations, much of it has focused on single-level models, such as linear regression. In the context of multilevel models, Lange, Little, and Taylor (1989) developed an approach for modeling data with outliers that involves the adjustment of random-effect distributions (e.g., ε_{ij} and U_{0j}) from the standard multivariate normal to heavy-tailed distributions, such as the multivariate t with υ degrees of freedom, or the Cauchy, as examples. Models based on such heavy-tailed random effects can better accommodate extreme values in the tails (i.e. outliers). In turn, such models should yield more accurate parameter and standard error estimates (Lange et al., 1989). Pinheiro et al. (2001) extended this work by describing a maximum likelihood (ML) algorithm for estimating multilevel model parameters using the multivariate t distribution. Research has shown that these heavy-tailed random-effects models yield more accurate parameter estimates and smaller standard errors than do models based on the assumption that errors are normally distributed (Song, Zhang, and Qu, 2007; Tong and Zhang, 2012; Yuan and Bentler, 1998; Yuan, Bentler, and Chan, 2004).

The standard multilevel model can be written as:

$$\left[y_i, U_{0j}\right] \sim N\left(\begin{bmatrix} X_i\beta \\ 0 \end{bmatrix}, \begin{bmatrix} Z_j\Psi Z'_j + \Lambda_j & Z_j\Psi \\ \Psi Z'_j & \Psi \end{bmatrix}\right) \tag{10.8}$$

where

β = vector of fixed effects
Λj = level-1 covariance matrix
Ψ = level-2 covariance matrix
X_i = design matrix for the fixed effects
Z_j = design matrix for the level-2 random effects

Pinheiro et al. (2001) showed that this model can be rewritten as:

$$\left[y_i, U_{0j}\right] \sim t\left(\begin{bmatrix} X_i\beta \\ 0 \end{bmatrix}, \begin{bmatrix} Z_j\Psi Z'_j + \Lambda_j & Z_j\Psi \\ \Psi Z'_j & \Psi \end{bmatrix}, v\right) \tag{10.9}$$

where

v = degrees of freedom for the t distribution

Given Equation (10.9), the dependent variable, y, follows the t distribution with v degrees of freedom, as do the level-1 and level-2 error terms, ε_{ij} and U_{0j} (Pinheiro et al. 2001). In turn, the variances of both ε_{ij} and U_{0j} are a function of the degrees of freedom. This last result means that there are is infinite number of possible distributions that the random-effects variances can follow, all based on t degrees of freedom.

The use of a heavy-tailed distribution to model the distributions of the random effects can be extended beyond the t with v degrees of freedom to include the Cauchy (Hogg and Tanis, 1996), and the slash (Rogers and Tukey, 1972) distributions. As with the t, these heavy-tailed distributions may be useful tools to account for the presence of outliers in the data. The Cauchy distribution has unknown mean and variance, but defined median and mode, and is symmetric with heavier tails than the normal distribution. The slash distribution is defined as the ratio of the normal (0,1) and uniform (0,1) distributions, and, like the t and Cauchy distributions, has been shown to be useful for parameter estimation in the presence of outliers (e.g. Wang and Genton, 2006). For the t and slash distributions, when v is unknown, as is most often the case, the expectation maximization (EM) algorithm that is used to estimate the model parameters includes an additional step in which the degrees of freedom are estimated as well. The logic underlying all of these approaches to parameter estimation when outliers are present is that the heavier-tailed distributions can better accommodate outliers than can the normal, resulting in lower parameter estimation bias and smaller standard errors (Welsh and Richardson, 1997).

A second alternative to ML/REML estimation of multilevel model parameters in the presence of outliers is based on a joint rank estimator (JR; Kloke et al., 2009). In JR, the raw scores of the dependent variable are replaced with their ranks based on a nondecreasing score function, such as the Wilcoxon (Wilcoxon, 1945). Assuming a common marginal distribution of level-1 errors (ε_{ij}) across level-2 units, estimation of the fixed effects (β_1, γ_{00}) is based on Jaekel's (1972) dispersion function:

$$\hat{\beta}_{\varphi} = \mathrm{Argmin} \left\| Y - X\hat{\beta} \right\|_{\varphi} \tag{10.10}$$

where
    Y = dependent variable
    X = matrix of independent variable values
    $\hat{\beta}$ = matrix of estimates of the fixed effects for the model

$$\left\| Y - X\hat{\beta}_{\varphi} \right\| = \sum_{i=1}^{N} \left[R\left(y_{ij} - \hat{y}_{ij}\right) \right] \left(y_{ij} - \hat{y}_{ij}\right)$$

R denotes the rank
    y_{ij} = dependent variable value for individual i in cluster j
    $\hat{y}_{ij}$ = model-based predicted dependent variable value for individual i in cluster j

In short, estimation of the fixed-effects parameters is based on a minimization of the residual ranks between the observed and predicted values of the dependent variable.

Estimation of fixed-effects parameter standard errors can be done using either of two approaches (Kloke et al., 2009). The first of these was based on the assumption that the within-cluster error terms are compound symmetric; i.e. a common covariance exists between any pair of observations. The primary advantage of this compound symmetric estimator (JR_CS) is that it is computationally very efficient, requiring estimation of only one additional parameter beyond those that are a part of the standard multilevel model, the common covariance between item pairs (Kloke et al., 2009). Its main limitation is the strong assumption of exchangeability of error terms, which may frequently not hold in actual practice. Given the potential problems associated with making this exchangeability assumption, Kloke and McKean (2013) proposed a sandwich estimator (JR_SE) for the model parameter standard errors. This JR_SE estimator does not require any additional assumptions about the data beyond those of ML/REML, unlike the case for JR_CS. Kloke and McKean (2013) conducted a simulation study comparing the performance of JR_CS and JR_SE, and found that JR_SE worked well for samples of 50 or more level-2 units. However, for fewer than 50 level-2 units, JR_SE yielded somewhat larger standard errors than was the case for JR_CS, thereby leading to more conservative inference with regard to the statistical significance of the parameter estimates. Kloke and McKean (2013) also found that when the exchangeability assumption was violated, JR_CS standard error estimates were inflated, also reducing power for inference regarding these parameters. Given this combination of results, Kloke and McKean recommended that JR_SE be used as the default method for estimating model parameter standard errors. However, when the level-2 sample size is small, researchers should use JR_CS, unless they know that the exchangeability assumption has been violated.

Fitting Robust and Rank-Based Multilevel Models in R

In order to demonstrate the fitting of robust and rank-based multilevel models in R, we will return to the dental data. Recall that there were four subjects who were flagged multiple times as potential outliers, and a fifth who was flagged once. In order to obtain the various outlier statistics, we had to fit a multilevel model, in this case with age and sex of the subject serving as independent variables, and distance from the center of the pituitary to the pterygomaxillary fissure. The R output with the parameter estimates appears below.

```
Random effects:
 Groups    Name         Variance Std.Dev.
 Subject   (Intercept)  3.267    1.807
 Residual               2.049    1.432
Number of obs: 108, groups:  Subject, 27

Fixed effects:
             Estimate Std. Error       df t value Pr(>|t|)
(Intercept) 17.70671    0.83392 99.35237  21.233  < 2e-16 ***
age          0.66019    0.06161 80.00000  10.716  < 2e-16 ***
SexFemale   -2.32102    0.76142 25.00000  -3.048  0.00538 **
---
Signif. codes:  0 '***' 0.001 '**' 0.01 '*' 0.05 '.' 0.1 ' ' 1
```

When we ignore the potential outliers, we would conclude that age has a statistically significant positive relationship with distance, and that, on average, females have smaller distance measures than do males. In addition, the variance component associated with subject is somewhat larger than that associated with error, indicating that there is a non-trivial degree of difference in distance measurements among individuals.

Now let's first fit this model using an approach based on ranks of the data, rather than the raw data itself. In order to do this, we will need to install the jrfit library from github. The commands for this installation appear below.

```
install.packages("devtools")
library(devtools)
install_github("kloke/jrfit")
```

We would then use the `library` command to load `jrfit`. After doing this, we will need to create a matrix (X) that contains the independent variables of interest, age, and sex.

```
library(jrfit)
X<-cbind(dental$age, dental$Sex)
```

We are now ready to fit the model. Recall that there are two approaches for estimating standard errors in the context of the rank-based approach, one based on an assumption that the covariance matrix of within-subject errors is compound symmetric, and the other using the sandwich estimator. We will employ both for this example. First, we will fit the model with the compound symmetry approach to standard error estimation.

```
Model10.2.cs<-jrfit(X,dental$distance,dental$Subject,var.typ
e='cs')
summary(model10.1.cs)

Coefficients:
    Estimate Std. Error t-value   p.value
X1 20.163737   1.421843 14.1814 < 2.2e-16 ***
X2  0.625000   0.063482  9.8454 < 2.2e-16 ***
X3 -2.163741   0.812548 -2.6629  0.008979 **
---
Signif. codes:  0 '***' 0.001 '**' 0.01 '*' 0.05 '.' 0.1 ' ' 1
```

Note that in these results, X1 denotes the intercept, X2 subject age, and X3 subject sex. These results are quite similar to those obtained using the standard multilevel model assuming that the data are normally distributed. The results for the sandwich estimator standard errors appear below.

```
Model10.2.sandwich<-jrfit(X,dental$distance,dental$Subject,v
ar.type='sandwich')
summary(model10.1.sandwich)

Coefficients:
    Estimate Std. Error t-value   p.value
X1 20.16374    1.64352 12.2686 1.493e-12 ***
X2  0.62500    0.07846  7.9658 1.461e-08 ***
X3 -2.16374    0.86223 -2.5095   0.01839 *
---
Signif. codes:  0 '***' 0.001 '**' 0.01 '*' 0.05 '.' 0.1 ' ' 1
```

The standard errors produced using the sandwich estimator were somewhat larger than those assuming compound symmetry, though in the final analysis the results with respect to relationships between the independent variables and the response were qualitatively the same. In summary, then, the standard errors yielded by the rank-based approach were somewhat larger than those produced by the standard model, though they did not differ substantially in this example. In addition, the results of the three models all yielded the same overall findings.

In order to fit the heavy-tailed estimators, we will need to use the heavy library in R. In order to fit the model based on the t distribution, the following command is used.

```
library(heavy)
Model10.3.t <- heavyLme(distance ~ age + Sex, groups = ~
Subject, data = dental, family = Student())
summary(model10.1.t)

Linear mixed-effects model under heavy-tailed distributions
 Data: dental; Family: Student(df = 5.36305)
 Log-likelihood: -212.9181

Random effects:
 Formula: ~age; Groups: ~Subject
 Scale matrix estimate:
             (Intercept) age
(Intercept)  3.66883654
age          -0.15707253  0.03418491
Within-Group scale parameter: 0.9017474

Fixed: distance ~ age + Sex
             Estimate Std.Error Z-value p-value
(Intercept)  17.6731    1.4737  11.9924  0.0000
age           0.6367    0.1194   5.3333  0.0000
SexFemale    -0.0740    1.7195  -0.0431  0.9657

Number of Observations: 108
Number of Groups: 27
```

The estimated degrees of freedom for the t that yielded the best fit was 5.36305. In terms of the fixed effects, the standard errors were larger than was the case for either the standard or rank-based models. Indeed, the standard error for sex was over twice as large for the robust t model as compared to either the rank or standard model estimates. The result of this larger standard error was a non-statistically significant test result for sex. In addition, the estimated coefficient for sex was also much smaller than the one yielded by the other two modeling approaches. Thus, based on this result, we would conclude that when children were older, they had larger distance measurements, but that there were not any such differences between males and females.

We can also fit this model using other heavy-tailed distributions, such as the Cauchy, the slash, and a contaminated normal. The commands and associated output for each of these models appear below.

Cauchy

```
Model10.4.slash <- heavyLme(distance ~ age + Sex, random = ~
age, groups = ~ Subject,data = dental, family = slash())
summary(model10.1.slash)
```

Linear mixed-effects model under heavy-tailed distributions
 Data: dental; Family: Cauchy()
 Log-likelihood: -221.8512

Random effects:
 Formula: ~age; Groups: ~Subject
 Scale matrix estimate:
            (Intercept) age
(Intercept) 2.97619934
age   -0.10957288 0.02658286
Within-Group scale parameter: 0.6567805

Fixed: distance ~ age + Sex
            Estimate Std.Error Z-value p-value
(Intercept) 18.0818   1.3993 12.9223 0.0000
age   0.6035   0.1110   5.4361 0.0000
SexFemale   -0.0522   1.7050 -0.0306 0.9756

Number of Observations: 108
Number of Groups: 27

Slash

```
Model10.5.slash <- heavyLme(distance ~ age + Sex, random = ~
age, groups = ~ Subject,data = dental, family = slash())
summary(model10.1.slash)
```

Linear mixed-effects model under heavy-tailed distributions
 Data: dental; Family: slash(df = 1.58618)
 Log-likelihood: -212.0925

Random effects:
 Formula: ~age; Groups: ~Subject
 Scale matrix estimate:
            (Intercept) age
(Intercept) 2.30753577
age   -0.08770319 0.02131623
Within-Group scale parameter: 0.5798421

Fixed: distance ~ age + Sex
            Estimate Std.Error Z-value p-value
(Intercept) 17.6585   1.4648 12.0551 0.0000
age   0.6420   0.1177   5.4539 0.0000
SexFemale   -0.0773   1.7474 -0.0442 0.9647
```
```

```
Number of Observations: 108
Number of Groups: 27
```

Contaminated

```
model10.1.contaminated <- heavyLme(distance ~ age + Sex,
random = ~ age, groups = ~ Subject,data = dental, family =
contaminated())
summary(model10.1.contaminated)

Linear mixed-effects model under heavy-tailed distributions
 Data: dental; Family: contaminated()
 Log-likelihood: -212.344

Random effects:
 Formula: ~age; Groups: ~Subject
 Scale matrix estimate:
             (Intercept) age
(Intercept)  3.53020881
age         -0.10475888  0.03117002
Within-Group scale parameter: 1.003603

Fixed: distance ~ age + Sex
             Estimate Std.Error Z-value p-value
(Intercept)  17.6482   1.4471   12.1957 0.0000
age           0.6503   0.1146    5.6757 0.0000
SexFemale    -0.0899   1.7443   -0.0515 0.9589

Number of Observations: 108
Number of Groups: 27
```

Though the estimates differ slightly across the estimation methods, they all yield the same basic result, which is that age is positively related to distance, and there are no differences between males and females.

Given these results, the reader will correctly ask the question, which of these methods should I use? There has not been a great deal of research comparing these various methods with one another. However, one study (Finch, 2017) did compare the rank-based, heavy-tailed, and standard multilevel models with one another, in the presence of outliers. The results of this simulation work showed that the rank-based approaches yielded the least biased parameter estimates, and had smaller standard errors than did the heavy-tailed approaches. Certainly, those standard error results are echoed in the current example, where the standard errors for the heavy-tailed approaches were more than twice the size of those from the rank-based method. Given these simulation results, it would seem that the rank-based results may be the best to use when outliers are present, at least until further empirical work demonstrates otherwise.

Multilevel Lasso

In some research contexts, the number of variables that can be measured (*p*) approaches, or even exceeds, the number of individuals on whom such measurements can be made (*N*). For example, researchers working with gene assays may have thousands of measurements that were made on a sample of only 10 or 20 individuals. The consequence of having such small samples coupled with a large number of measurements is known as high-dimensional data. In such cases, standard statistical models often do not work well, yielding biased standard errors for the model parameter estimates (Bühlmann and van de Geer, 2011). These biased standard errors in turn can lead to inaccurate Type I error and power rates for inferences made about these parameters. High dimensionality can also result in parameter estimation bias due to the presence of collinearity (Fox, 2016). Finally, when *p* exceeds *N*, it may not be possible to obtain parameter estimates at all using standard estimators.

In the context of standard single-level data structures, statisticians have worked to develop estimation methods that can be used with high-dimensional data, with one of the more widely used such approaches being regularization or shrinkage methods. Regularization methods involve the application of a penalty to the standard estimator such that the coefficients linking the independent variables to the dependent variables are made smaller, or shrunken. The goal of this technique is that only those variables that are most strongly related to the dependent variable are retained in the model, whereas the others are eliminated by having their coefficients reduced (shrunken) to 0. This approach should eliminate from the model independent variables that exhibit weak relationships to the dependent variable, thereby rendering a reduced model.

One of the most popular regularization approaches is the least absolute shrinkage and selection operator (lasso; Tibshirani, 1996), which can be expressed as

$$e^2 = \sum_{i=1}^{N} \left(y_i - \hat{y}_i\right)^2 + \lambda \sum_{j=1}^{p} \left\lfloor \hat{\beta}_j \right\rfloor. \tag{10.11}$$

where

y_i = the observed value of the dependent variable for individual *i*
$\hat{y}_i$ = the model-predicted value of the dependent variable for individual *i*
$\hat{\beta}_j$ = sample estimate of the coefficient for independent variable *j*
λ = shrinkage penalty-tuning parameter

The tuning parameter, λ, is used to control the amount of shrinkage, with larger λ values corresponding to greater shrinkage of the model; i.e. a greater reduction in the number of independent variables that are likely to be included in the final model. If λ is set to 0, the resulting estimates are equivalent to those produced by the standard estimator; i.e. least squares, ML.

In summary, the goal of the lasso estimator is to eliminate from the model those independent variables that contribute very little to the explanation of the dependent variable, by setting their $\hat{\beta}$ values to 0, while at the same time retaining independent variables that are important in explaining y. The optimal λ value is specific to each data analysis problem. A number of approaches for identifying it have been recommended, including the use of cross-validation to minimize the mean squared error (Tibshirani, 1996), or selection of λ that minimizes the Bayesian information criterion (BIC). This latter approach was recommended by Schelldorfer, Bühlmann, and van de Geer (2011), who showed that it works well in many cases. Zhao and Yu (2006) also found the use of the BIC for this purpose to be quite effective. With this approach, several values of λ are used, and the BIC values for the models are compared. The model with the smallest BIC is then selected as being optimal.

Schelldorfer et al. (2011) described an extension of the lasso estimator that can be applied to multilevel models. The multilevel lasso (MLL) utilizes the lasso penalty function, with additional terms to account for the variance components associated with multilevel models. The MLL estimator minimizes the following function:

$$Q_\lambda\left(\beta,\tau^2,\sigma^2\right):=\frac{1}{2}\ln|V|+\frac{1}{2}\left(y_i-\hat{y}_i\right)'V^{-1}(y_i-\hat{y}_i)+\lambda\sum_{j=1}^{p}\left\lfloor\hat{\beta}_j\right\rfloor \qquad (10.12)$$

where
τ^2 = between cluster variance at level 2
σ^2 = within cluster variance at level 1
V = covariance matrix

From Equation (10.12), we can see that model parameter estimates are obtained with respect to penalization of level-1 coefficients, and otherwise work similarly to the single-level lasso estimator. In order to conduct inference for the MLL model parameters, standard errors must be estimated. However, the MLL algorithm currently does not provide standard error estimates, meaning that inference is not possible. Therefore, interpretation of results from analyses using MLL will focus on which coefficients are not shrunken to 0, as we will see in the example below.

Fitting the Multilevel Lasso in R

In order to demonstrate the use of the multilevel lasso model, we will use the `classroomStudy` data that is part of the `lmmlasso` R library. In this example, the gain in student math achievement scores will serve as the dependent variable, with the independent variables including student sex, minority status, type of math instruction, student socioeconomic status, years of teacher experience, and time spent in math preparation. The level-2 effect was classroom. The level-1 sample size was 156 students, with level 2 having 44 classrooms. First, we will fit this model employing the standard multilevel model, using lme4.

```
library(lmmlasso)
data(classroomStudy)
Model10.6<-lmer(classroomStudy$y~classroomStudy$X+(1|classroom
Study$grp), data=classroomStudy)
summary(Model10.6)
```

The syntax that we use here generally matches the general structure demonstrated in Chapter 3. Notice that the major difference from the examples in the earlier chapter is that the independent fixed-effects variables are collected in the matrix X. The resulting output for the fixed and random effects appears below.

```
Random effects:
 Groups                 Name          Variance Std.Dev.
 classroomStudy$grp  (Intercept)  0.1583    0.3979
 Residual                            0.4745    0.6888
Number of obs: 156, groups:  classroomStudy$grp, 44

Fixed effects:
                            Estimate Std. Error        df t
value Pr(>|t|)
(Intercept)                  0.02599    0.08611   26.90650
0.302     0.765
classroomStudy$Xsex         -0.07690    0.06046  141.64555
-1.272     0.206
classroomStudy$Xminority    -0.02015    0.06803  134.13265
-0.296     0.768
classroomStudy$Xmathkind    -0.63959    0.06356  148.47809
-10.063    <2e-16 ***
classroomStudy$Xses          0.05329    0.06119  143.19103
0.871     0.385
classroomStudy$Xyearstea    -0.01861    0.09656   28.41109
-0.193     0.849
classroomStudy$Xmathprep    -0.06041    0.10045   24.26140
-0.601     0.553
---
Signif. codes:  0 '***' 0.001 '**' 0.01 '*' 0.05 '.' 0.1 ' ' 1
```

From these results, we would conclude that the only statistically significant effect was kind of math instruction, with a coefficient of −0.64.

Now, let's fit the same model using the multilevel lasso. In order to do this, we will need to install and then load the lmmlasso package in R, which we have done above. The R commands to fit the model, and to obtain a summary of the output, appear below.

```
Model10.7 <-lmmlasso(x=classroomStudy$X,y=classroomStudy$y,z
=classroomStudy$Z,grp=classroomStudy$grp,lambda=15,pdMat="pd
Ident")
summary(Model10.7)
```

There are several things to note when using lmmlasso. First, the fixed-effects variables are collected in the matrix X, which we have already commented on when discussing the use of lmer with the classroomStudy data. When using lmmlasso, the fixed effects will always need to be in such a matrix. Second, the object classroomStudy$Z is a column of 1s in this case, indicating that we have only a random slope. If we were also fitting random slopes for one or more independent variables, then those variables would need to each have a column in the Z matrix. The level-2 effect itself, grp, appears after the grp subcommand. We must set the value of lambda, which in this case is 15. We will try other values, and compare model fit using AIC and BIC, selecting the lambda that minimizes these values. Finally, the pdMat subcommand defines the covariance structure for the random effect, in this case setting it to be the identity matrix.

The output for Model10.7 appears below.

```
Model fitted by ML for lambda = 15 :
     AIC       BIC     logLik  deviance objective
   363.9     379.1     -176.9     353.9     186.0

Random effects: pdIdent
              Variance   Std.Dev.
Intercept  0.1227391  0.3503414
Residual   0.4769248  0.6905974

Fixed effects:
|active set|= 3
                 Estimate
(Intercept)   0.02065969 (n)
sex          -0.02376197
mathkind     -0.57887176

Number of iterations: 4
```

Using a lambda of 15, only two fixed effects had non-zero coefficients, sex and mathkind. Also, notice that the coefficients for these effects are smaller than were those estimated using the standard multilevel model, reflecting the shrinkage associated with the lasso. Let's now fit a model with a lambda of 10.

```
Model10.8 <-lmmlasso(x=classroomStudy$X,y=classroomStudy$y,z
=classroomStudy$Z,grp=classroomStudy$grp,lambda=10,pdMat="pd
Ident")
summary(Model10.8)
```

```
Model fitted by ML for lambda = 10 :
    AIC        BIC     logLik  deviance objective
   364.8      383.1    -176.4     352.8     182.9

Random effects: pdIdent
              Variance  Std.Dev.
Intercept 0.1211003 0.3479947
Residual  0.4742756 0.6886767

Fixed effects:
|active set|= 4
                  Estimate
(Intercept)   0.02039525 (n)
sex          -0.04084089
mathkind     -0.59798409
ses           0.00546273

Number of iterations: 5
```

We can see that there is less shrinkage in the fixed-effects coefficients, which we expect given that the penalty term for this model is smaller. However, also notice that the AIC and BIC values are larger for Model10.8 than was the case for Model10.7, leading us to conclude that we may need a larger penalty term. Next, let's set lambda at 20 and see what happens to the information indices.

```
Model10.9 <-lmmlasso(x=classroomStudy$X,y=classroomStudy$y,z
=classroomStudy$Z,grp=classroomStudy$grp,lambda=20,pdMat="pd
Ident")
summary(Model10.9)

Model fitted by ML for lambda = 20 :
    AIC        BIC     logLik  deviance objective
   365.1      380.4    -177.6     355.1     188.9

Random effects: pdIdent
              Variance  Std.Dev.
Intercept 0.1247968 0.3532659
Residual  0.4802174 0.6929772

Fixed effects:
|active set|= 3
                  Estimate
(Intercept)   0.021169801 (n)
sex          -0.006581567
mathkind     -0.560115122

Number of iterations: 5
```

The fixed-effects parameter estimate shrinkage is greater when lambda is 20, as we would anticipate. The AIC and BIC values are larger than those for a lambda of 15, indicating that the latter is a preferable value in terms of model fit. We could repeat this sequence of model fitting and comparison of information indices for various values of lambda in order to obtain what we believe to be the optimal setting. Given what we have already seen here, it is most likely that this optimal lambda value is close to 15, and so we would concentrate our exploration in that neighborhood.

Multivariate Multilevel Models

In some cases, a researcher will have collected data on more than one outcome variable for each member of the sample. For example, in an educational research study focused on identifying predictors of student achievement, test scores for math, reading, and science might be collected from students, as well as a measure of socioeconomic status (SES). In this context, the research question of interest would revolve around simultaneously examining relationships between SES and the dependent variables. In a traditional univariate approach to this problem, the data analyst would likely model the relationship between the predictor, SES, and each of the dependent variables separately, providing information about the relationships of SES with each response separately. Two problems would emerge from using this multiple univariate modeling approach, however. First, the overall Type I error rate would be inflated because separate models were being used with the same data and same predictor(s). This problem is easily dealt with using some type of correction to the level of α, such as Bonferroni. The second issue with this multiple modeling approach is that the correlation structure among the dependent variables is ignored when model parameters are estimated. Indeed, the correlation coefficients are assumed to be 0 for all pairs of dependent variables. Clearly in many, perhaps most, real-world applications such a limited correlation structure is not realistic. Therefore, we need to develop a multivariate modeling approach. Such models are widely used in single-level modeling in the form of multivariate analysis of variance (MANOVA) and multivariate regression.

Although work on multivariate multilevel modeling has been less widely explored than is the case for univariate multilevel models, there are extant techniques for fitting such models using R. In particular, Snijders and Bosker (2012) proposed just such an approach, and then demonstrated how to fit the model using R. For the case with p dependent variables and one predictor, their model takes the form:

$$y_{icp} = \gamma_{0p} + \gamma_{1p} x_{1ic} + U_{pc} + R_{icp} \tag{10.13}$$

where

y_{icp} = value on dependent variable p for individual i in cluster c
γ_{0p} = intercept for variable p
x_{1ic} = value on independent variable 1 for individual i in cluster c
γ_{1p} = coefficient linking independent variable 1 to dependent variable p
U_{pc} = random error component for cluster c on dependent variable p
R_{icp} = random error for dependent variable p for individual i in cluster c

Given that there are p dependent variables in model (10.13), the random-effect variances in a univariate multilevel model (i.e. variances of U_c and R_{ic}) become the covariance matrices for these model terms:

$$\Sigma = \text{cov}(R_{ic}) \tag{10.14}$$

$$T = \text{cov}(U_c)$$

Conceptually, Model 10.13 allows for the assessment of relationships between the predictor variable (x_1) and each of the dependent variables (y_p), while accounting for the correlations among the dependent variables. In order to fit this model, the data must be reconfigured so that each individual in the sample has a separate line for each of the p dependent variables. In addition, the dependent variables must be represented by a set of p dummy variables, $d_1, d_2, \ldots, d_p$. The value of $d_p=1$ for the line of data containing the dependent variable p, and 0 for this variable otherwise. Using these dummy variables, Equation (10.13) can be reexpressed as a three-level model in which level 1 corresponds to the dependent variable, level 2 to the individual member of the sample, and level 3 to the clustering variable (e.g. school), as in Equation (10.13).

$$y_{icp} = \sum_{s=1}^{m} \gamma_{0sp} d_{sicp} + \sum_{p=1}^{k}\sum_{s=1}^{m} \gamma_{1sp} x_{1sic} d_{sicp} + \sum_{p=1}^{k}\sum_{s=1}^{m} U_{spc} d_{sicp}$$

$$+ \sum_{p=1}^{k}\sum_{s=1}^{m} R_{sicp} d_{sicp} \tag{10.15}$$

This model yields hypothesis testing results for each of the response variables, accounting for the presence of the others in the data. In order to test the multivariate null hypothesis of group mean equality across all of the response variables, we can fit a null multivariate model to the data for which the independent variables are not included. The fit of this null model can then be compared with that of the full model including the independent variable(s) of interest to test the null hypothesis of no multivariate group mean differences, if the independent variable is categorical. This comparison can be carried out using a likelihood ratio test. If the resulting p-value is below the threshold (e.g. $\alpha = 0.05$) then we would reject the null hypothesis of no group mean differences, because the fit of the full model including the group effect was better than that of the null model. The reader interested in applying this model using R is encouraged to read the excellent discussion and example provided in Chapter 16 of Snijders and Bosker (2012).

Multilevel Generalized Additive Models

Generalized additive models (GAMs) link an outcome variable, y, with one or more independent variables, x, using smoothing splines. Splines are piecewise polynomials for which individual functional sections are joined at locations in the data known as knots (Hastie and Tibshirani, 1990). A commonly used such approach is the cubic spline, which is simply a set of cubic polynomial functions joined together at the knots, for which each section between the knots has unique model parameter values. It appears in (10.16).

$$y = \beta_0 + \beta_1 x + \beta_2 x^2 + \beta_3 x^3 \tag{10.16}$$

where
  y = dependent variable
  x = independent variable
  β_j = coefficient for model term j

The cubic spline then fits different versions of this model between each pair of adjacent knots. The more knots in the GAM, the more piecewise polynomials will be estimated, and the more potential detail about the relationship between x and y will be revealed. GAMs take these splines and apply them to a set of one or more predictor variables as in (10.17).

$$y_i = \beta_0 + \Sigma f_j(x_i) + \varepsilon_i \tag{10.17}$$

where
  y_i = value of outcome variable for subject i
  β_0 = intercept
  f_j = smoothing spline for independent variable j, such as a cubic spline
  x_i = value of independent variable for observation i
  ε_i = random error, distributed $N(0, \sigma^2)$

Each independent variable has a unique smoothing function, and the optimal set of smoothing functions is found by minimizing the penalized sum of squares criterion (PSS) in (10.18).

$$\text{PSS} = \sum_{i=1}^{N} \left\{ y_i - \beta_0 + \sum f_j(x_i) \right\}^2 + \sum_{j=1}^{p} \lambda_j \int f_j''(t_j)^2 dt_j \qquad (10.18)$$

Here, y_i is the value of the response variable for subject i, and λ_j is a tuning parameter for variable j such that $\lambda_j \geq 0$. The researcher can use λ_j to control the degree of smoothing that is applied to the model. A value of 0 results in an unpenalized function and relatively less smoothing, whereas values approaching ∞ result in an extremely smoothed (i.e. linear) function relating the outcome and the predictors. The GAM algorithm works in an iterative fashion, beginning with the setting of β_0 to the mean of Y. Subsequently, a smoothing function is applied to each of the independent variables in turn, minimizing the PSS. The iterative process continues until the smoothing functions for the various predictor variables stabilize, at which point final model parameter estimates are obtained. Based upon empirical research, a recommended value for λ_j is 1.4 (Wood, 2006), and as such will be used in this study.

GAM rests on the assumption that the model errors are independent of one another. However, when data are sampled from a clustered design, such as measurements made longitudinally for the same individual, this assumption is unlikely to hold, resulting in estimation problems, particularly with respect to the standard error (Wang, 1998). In such situations, an alternative approach to modeling the data, which accounts for the clustering of data points, is necessary. The generalized additive mixed model (GAMM) accounts for the presence of clustering in the form of random effects (Wang, 1998). GAMM takes the form:

$$y_i = \beta_0 + \Sigma f_j(x_i) + Z_i b + \varepsilon_i \qquad (10.19)$$

where
   Z_i = random effects in the model; e.g. person being measured
   b = random-effects coefficients

Other model terms in (4) are the same as in (2). GAMM is fit in the same fashion as GAM, with an effort to minimize PSS, and the use of λ_j to control the degree of smoothing.

Fitting GAMM using R

We can fit a GAMM using functions from the gamm4 library. As an example, let's return to the achievement data that we examined in Chapter 3. Specifically, we are interested in the random intercept model in which reading score (geread) is the dependent variable, and the verbal score (npaverb) is the predictor. In this case, however, we would use splines to investigate whether there exists a nonlinear relationship between the two variables. The R code for fitting this model appears below.

```
Model10.8<-gamm4(geread~s(npaverb), family=gaussian,
random=~(1|school), data=prime_time)
```

The structure of the model fitting commands is quite similar to what we have seen heretofore with lme4. The primary function used in fitting this model is gamm4, and the model syntax poses the dependent variable across ~ from the independent variable. The primary difference here is that rather than npaverb itself, the predictor is a smoother involving npaverb, s(npaverb). By using family=gaussian, we are assuming the normality of the model errors. Finally, the random effect is expressed much as is the case with lme4, with 1|school indicating that we are fitting a random intercept model, with school as the clustering variable.

The output from gamm4 is divided into the multilevel portion (mer) and the GAM portion (gam). We can see the results for each of these using the summary command, as below.

```
summary(Model10.8$mer)

Linear mixed model fit by REML ['lmerMod']

REML criterion at convergence: 45286.1

Scaled residuals:
    Min      1Q  Median      3Q     Max
-2.3364 -0.6007 -0.1976  0.3110  4.7359

Random effects:
 Groups   Name         Variance Std.Dev.
 school   (Intercept)  0.1032   0.3213
 Xr       s(npaverb)   2.0885   1.4452
 Residual              3.8646   1.9659
Number of obs: 10765, groups:  school, 163; Xr, 8

Fixed effects:
                 Estimate Std. Error t value
X(Intercept)      4.34067    0.03215  135.00
Xs(npaverb)Fx1    2.30471    0.25779    8.94

Correlation of Fixed Effects:
            X(Int)
Xs(npvrb)F1 -0.002
```

Here we see estimates for variance associated with the school random effect, along with the residual. The smoother also has a random component, though it is not the same as a random slope for the linear portion of the model, which we will see below. The fixed effects in this portion of the output correspond to those that we would see in a linear model, so that X(Intercept) is a standard intercept term, and Xs(npaverb)Fx1 is the estimate of a linear relationship between npaverb and geread. Here we see that the *t*-value for this linear effect is 8.94, suggesting that there may be a significant linear relationship between the two variables.

The nonlinear smoothed portion of the relationship can be obtained using the following command.

```
summary(Model10.8$gam)

Family: gaussian
Link function: identity

Formula:
geread ~ s(npaverb)

Parametric coefficients:
            Estimate Std. Error t value Pr(>|t|)
(Intercept)  4.34067    0.03215     135   <2e-16 ***
---
Signif. codes:  0 '***' 0.001 '**' 0.01 '*' 0.05 '.' 0.1 ' ' 1

Approximate significance of smooth terms:
             edf Ref.df      F p-value
s(npaverb) 6.031  6.031  609.4  <2e-16 ***
---
Signif. codes:  0 '***' 0.001 '**' 0.01 '*' 0.05 '.' 0.1 ' ' 1

R-sq.(adj) =    0.28
lmer.REML =  45286   Scale est. = 3.8646     n = 10765
```

When reading these results, note that the values for the intercept fixed effect and its standard error are identical to the results from the mer output. Of more interest when using GAM is the nature of the smoothed nonlinear term, which appears as s(npaverb). We can see that this term is statistically significant, with a *p*-value well below 0.05, meaning that there is a nonlinear relationship between npaverb and geread. Also note that the adjusted R^2 is 0.28, meaning that approximately 28% of the variance in the reading score is associated with the model. In order to characterize the nature of the relationship between the reading and verbal scores, we can examine a plot of the GAM function.

```
plot(Model10.8$gam)
```

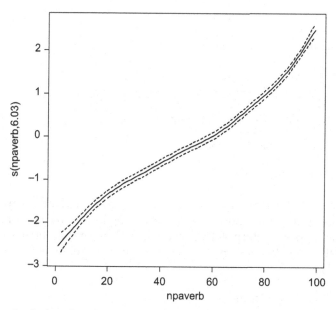

The plot includes the estimated smoothed relationship between the two variables, represented by the solid line, and the 95% confidence interval of the curve, which appears as the dashed lines. The relationship between the two variables is positive, such that higher verbal scores are associated with higher reading scores. However, this relationship is not strictly linear, as we see a stronger relationship between the two variables for those with relatively low verbal scores, as well as those for relatively higher verbal scores, when compared with individuals whose verbal scores lie in the midrange.

Just as with linear multilevel models, it is also possible to include random coefficients for the independent variables. The syntax for doing so, along with the resulting output, appears below.

```
Model10.9<-gamm4(geread~s(npaverb), family=gaussian, random=~(
npaverb|school), data=prime_time)

summary(Model10.9$mer)
Linear mixed model fit by REML ['lmerMod']

REML criterion at convergence: 45167.4

Scaled residuals:
    Min      1Q  Median      3Q     Max
-2.8114 -0.5995 -0.2001  0.3159  4.8829

Random effects:
 Groups   Name        Variance  Std.Dev. Corr
 school   (Intercept) 2.414e-02 0.1554
          npaverb     8.101e-05 0.0090   -1.00
```

```
 Xr           s(npaverb)    1.606e+00 1.2672
 Residual                   3.805e+00 1.9507
Number of obs: 10765, groups:  school, 163; Xr, 8

Fixed effects:
                Estimate Std. Error t value
X(Intercept)      4.32243    0.03308 130.671
Xs(npaverb)Fx1    2.18184    0.24110   9.049

Correlation of Fixed Effects:
            X(Int)
Xs(npvrb)F1 0.069
```

The variance for the random effect of npaverb is smaller than are the variance terms for the other model terms, indicating its relative lack of import in this case. We would interpret this result as indicating that the linear portion of the relationship between npaverb and geread is relatively similar across the schools.

```
summary(Model10.9$gam)

Family: gaussian
Link function: identity

Formula:
geread ~ s(npaverb)

Parametric coefficients:
            Estimate Std. Error t value Pr(>|t|)
(Intercept)  4.32243    0.03308   130.7   <2e-16 ***
---
Signif. codes:  0 '***' 0.001 '**' 0.01 '*' 0.05 '.' 0.1 ' ' 1

Approximate significance of smooth terms:
             edf Ref.df      F p-value
s(npaverb) 5.742  5.742  327.8  <2e-16 ***
---
Signif. codes:  0 '***' 0.001 '**' 0.01 '*' 0.05 '.' 0.1 ' ' 1

R-sq.(adj) =   0.28
lmer.REML =  45167  Scale est. = 3.8052     n = 10765
```

The smoothed component of the model is the same for the random coefficients model as it was for the random intercept-only model. This latter result is further demonstrated in the plot below. It appears to be very similar to that for the non-random coefficient model.

```
plot(Model10.9$gam)
```

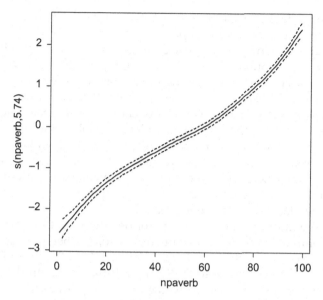

GAMs are powerful tools for estimating nonlinear relationships among variables. As was evidenced here, they also extend very easily into the multilevel modeling framework. We elected to use the gamm4 library, but the reader should also be aware of the mgcv library, which can also be used to fit multilevel GAM models. In either case, our focus is typically on the smoothed portion of the model, as it provides us with information about the nonlinear relationships among the predictor and dependent variables, assuming that such exist. It is also worth noting here that should the relationship be only linear in nature, this will be reflected by the GAM in the form of a strictly linear smoothed function, evidenced in the plot. Finally, there are a large number of smoothing parameters that we can use to improve the fit of the model, and hone in on a more accurate picture of the relationships among our variables. For the examples appearing in this chapter, we used the default settings for gamm4, but you should be aware that in practice it may be beneficial to examine a number of different settings in order to obtain a more accurate picture of the true model structure.

Predicting Level-2 Outcomes with Level-1 Variables

Throughout this book, when we have examined the relationship between level-1 and level-2 variables, those at level 1 have always served as the response, whereas the level-2 variables have been the predictors (see the examples in Chapter 3). For example, we may be interested in the relationship between the proportion of children in a school who receive free/reduced lunch (independent variable) and the performance of individual students on a test of academic achievement (dependent variable). Such cross-level models are very common. However, in some cases, we may be interested in examining the nature of relationships that move in the other direction. As an example, consider a situation in which the researcher is interested in examining the relationship between individual employee satisfaction scores and yearly profits for a company. In this case, the dependent variable is profit, collected once for each firm, whereas the independent variable is the employee satisfaction score, which is collected once for each individual working within the companies from which the data were gathered. These data follow the multilevel structure, where employees are nested in company. However, unlike examples that we have used heretofore, the independent variable is at level 1 and the dependent variable is at level 2.

Perhaps the most common approach for dealing with this data structure, which is commonly referred to as micro-macro, has been to aggregate the level-1 variable values into a single score for each level-2 unit, and then to fit a standard regression model to the data (Hoffman, 2002). In the aforementioned example, this would entail averaging the employee satisfaction scores so that each company had only one value, and then fitting a regression model where the yearly profits are predicted by this aggregated satisfaction score. This approach, though simple to execute, will yield negatively biased standard errors, as the averaging of values yields underestimates of the variance in the level-1 scores (Croon and van Veldhoven, 2007). Thus, an alternative approach for dealing with such data structure is necessary. Croon and van Veldhoven proposed just such a methodology, which is based on reframing the problem as one involving a level-2 latent independent variable with the level-1 scores serving as its indicators.

The micro-macro model proposed by Croon and van Veldhoven (2007) is built on a structural equation modeling framework, in which the level-1 observations are linked together into a latent variable at level 2. This latent level-2 predictor is then regressed onto the level-2 outcome variable. These two aspects of the micro-macro model are expressed below.

$$x_{ig} = \xi_g + \upsilon_{ig} \tag{10.20}$$

where
x_{ig} = value of independent variable for individual i in group g
ξ_g = group-level score on the independent variable
υ_{ig} = error term for individual i in group g

The relationship between the level-2 dependent variable and the latent level-2 predictor variable can then be written as

$$y_g = \beta_0 + \beta_1\xi_g + \epsilon_g \tag{10.21}$$

where

y_g = value of the dependent variable for group g

ϵ_g = error term associated with group g

β_1 = coefficient linking the latent predictor with the level-2 outcome

β_0 = intercept

Several assumptions underlie this modeling approach:

1. σ_ϵ^2, the variance of ϵ, is constant for all groups.
2. $\sigma_{v_{ig}}^2$, the variance of v, is constant for all individuals and groups.
3. v and ϵ are independent of one another and of ξ.
4. Level-1 units in the same group are assumed to have the same relationship with ξ.

In the estimation of this model, the best linear unbiased predictors (BLUPs) of the group means must be used in the regression model, rather than the standard unadjusted means, in order for the regression coefficient estimates to be unbiased. Croon and van Veldhoven provide a detailed description of how this model can be fit as a structural equation model. We will not delve into these details here, but do refer the interested reader to this article. Of key import here is that we see the possibility of linking these individual-level variables with a group-level outcome variable, using the model described above.

The micro-macro model can be fit in R using the `MicroMacroMultilevel` library. The following R code demonstrates first how the variables are appropriately structured for use with the library, followed by how the model is fit and subsequently summarized.

```
#REMOVE MISSING DATA#
prime_time.nomiss<-na.omit(prime_time)

#CREATE LEVEL-2 ID AND PREDICTOR VARIABLES#
z.gid<-as.vector(tapply(prime_time.nomiss$school, prime_time.
nomiss$school, mean))
z.mean<-as.vector(tapply(prime_time.nomiss$sattend, prime_
time.nomiss$school, mean))
z.data<-data.frame(z.gid, z.mean)

#CREATE LEVEL-1 ID AND PREDICTOR VARIABLES#
x.gid<-as.vector(prime_time.nomiss$school)
x.data<-data.frame(prime_time.nomiss$gemath, prime_time.
nomiss$gelang)

#CREATE LEVEL-2 RESPONSE VARIABLE#
y.prime_time<-as.vector(tapply(prime_time.nomiss$avgcsi1,
prime_time.nomiss$school, mean))
```

```
#OBTAIN ADJUSTED PREDICTOR MEANS#
micromacro.results.primetime<-adjusted.predictors(x.data,
z.data, x.gid, z.gid)

#SPECIFY MODEL FORMULA#
model.formula <- as.formula(y.prime_time ~ BLUP.prime_time.
nomiss.gemath + BLUP.prime_time.nomiss.gelang + z.mean)

#FIT MODEL AND OBTAIN SUMMARY OF RESULTS#
model.output = micromacro.lm(model.formula, micromacro.resu
lts.primetime$adjusted.group.means, y.prime_time, micromacro
.results.primetime$unequal.groups)

micromacro.summary(model.output)
```

In order to use the `MicroMacroMultilevel` library commands, missing values must be removed, and the data must be structured in a very specific way, as demonstrated above. Once this is done, the BLUP-adjusted means can be obtained using the `adjusted.predictors` command. These are then used as independent variables in conjunction with the level-2 predictor in the regression model. In the current example, the dependent variable is the school mean score on a cognitive skills index (avgcsi1), the level-2 predictor is the average daily attendance (sattend), and the level-1 predictors are student scores on the math and language exams (gemath and gelang). Note that the z.data object must include the level-2 identifier (z.gid) and values of the level-2 predictor(s). The results of the analysis appear below.

```
Call:
micromacro.lm( y.prime_time ~ BLUP.prime_time.nomiss.gemath +
BLUP.prime_time.nomiss.gelang + z.mean, ...)

Residuals:
      Min         1Q        Median          3Q       Max
 -13.25487  -2.442717  -0.0003431752  2.850811  12.2095
```

```
Coefficients:
                                     Estimate   Uncorrected S.E.
Corrected S.E.    df           t
(Intercept)                       -190.52742279    30.98161003
28.24900466 156 -6.7445712
BLUP.prime_time.nomiss.gemath       -0.01019005     0.02838026
0.02303021 156 -0.4424645
BLUP.prime_time.nomiss.gelang        0.01604510     0.02405509
0.01940692 156  0.8267722
z.mean                               3.04344765     0.32214229
0.29422655 156 10.3438919
```

```
                                     Pr(>|t|)              r
(Intercept)                       2.824611e-10 0.47514746
BLUP.prime_time.nomiss.gemath 6.587660e-01 0.03540331
BLUP.prime_time.nomiss.gelang 4.096290e-01 0.06605020
z.mean                            2.012448e-19 0.63783643

---
Residual standard error: 4.29361 on 156 degrees of freedom
Multiple R-squared: 0.3649649621, Adjusted R-squared:
0.3527527498
F-statistic: 29.88525 on 3 and 156 DF, p-value: 0
```

From these results, we can see that neither level-1 independent variable was statistically significantly related to the dependent variable, whereas the level-2 predictor, average daily attendance, was positively associated with the school mean cognitive skills index score. The total variance in the cognitive index that was explained by the model was 0.353, or 35.3%.

Power Analysis for Multilevel Models

In planning research, researchers may want to conduct a power analysis before gathering data, in order to determine how large their sample should be so as to have sufficient power for detecting the effect(s) of interest. Power is the probability of detecting an effect based on our sample data, when said effect is present in the population. For example, we may be interested in estimating the relationship between two variables, so that power is the probability of identifying a statistically significant relationship in the sample when the relationship in the population is not 0. In this context, a power analysis might be conducted prior to the beginning of the study in order to determine how large the sample size should be to achieve some acceptable level of power, such as 0.8.

For some statistics, such as the F used in analysis of variance, power can be calculated analytically, given the expected size of the effect, the sample size, and the desired level of α (e.g. 0.05). However, for more complex statistics and analyses, including multilevel models, such analytic solutions for calculating power are not available, or are difficult to use. In such instances, computer simulations can be used to provide information about power, and to determine, a priori, what the sample size should be for a given research scenario. In order to conduct such simulations, we need to have some ideas regarding the nature of the model that we will be fitting, and how large the effect(s) of interest are likely to be. For example, we may know that our model will include two level-1 independent variables, and a level-1 dependent variable. In addition, based on prior research in the field, we may anticipate an intraclass correlation (ICC) of approximately 0.2. The effects of primary interest in this scenario are the coefficients linking each of the independent variables to the dependent variable. In order to properly conduct a power analysis, we will also need to have some idea as to the expected magnitudes of these coefficients. These expectations can come from prior work in the field, or from our expectations based on previous observations. With this information in hand, we can construct statistical simulations that will provide us with information about power, as well as the quality of the parameter estimates. Such simulations can be conducted using the simr package in R.

In order to use the simr package for conducting a power analysis, we will need to specify the various parameters associated with the multilevel model of interest, including the sample sizes at levels 1 and 2, the magnitude of the fixed effects (intercepts and slopes), the variances for the random effects, and the covariance matrix of the random effects. Continuing with the example begun above, we have one independent variable, a single outcome, and an ICC of 0.2. Let's start with a level-1 sample size of 500, and a level-2 sample size of 25. The following specifications in R will produce the parameters for such a data structure.

```
library(simr)

x1 <- 1:20
cluster <- letters[1:25]

sim_data <- expand.grid(x1=x1, cluster=cluster)

b <- c(1, 0.5) # fixed intercept and slope
V1 <- 0.25 # random intercept variance
V2 <- matrix(c(0.5,0.3,0.3,0.1), 2) # random effects variance-
covariance matrix
s <- 1 # residual standard deviation

sim_model1 <- makeLmer(y ~ x1 + (1|cluster), fixef=b,
VarCorr=V1, sigma=s, data=sim_data)
sim_model1
```

The preceding code specifies the level-1 (x1) and level-2 (cluster) sample sizes, and places them in the object sim _ data using the expand.grid function, which is part of base R. We then specify the fixed effects for the intercept (1) and the slope (0.5) for our particular problem. The variance of the random intercept term is then set at 0.25 for this example, and the covariance matrix of the random effects is specified in the object V2. The standard deviation for the error term is 1. Finally, the makeLmer command is used to generate our model, which appears below. This output is useful for checking that everything we think we specified is actually specified correctly, which is the case here.

```
Linear mixed model fit by REML ['lmerMod']
Formula: y ~ x1 + (1 | cluster)
   Data: sim_data
REML criterion at convergence: 1479.115
Random effects:
 Groups    Name         Std.Dev.
 cluster   (Intercept)  0.5
 Residual               1.0
Number of obs: 500, groups:  cluster, 25
Fixed Effects:
(Intercept)            x1
        1.0           0.5
```

In order to generate the simulated data and obtain an estimate of the power for detecting the coefficient of 0.5 given the sample size parameters we have put in place, we will use the powerSim command from the simr library. We must specify the model and the number of simulations that we would like to use. In this case, we request 100 simulations. The results of these simulations appear below.

```
powerSim(sim_model1, nsim=100)

Power for predictor 'x1', (95% confidence interval):
      100.0% (96.38, 100.0)

Test: Kenward Roger (package pbkrtest)
      Effect size for x1 is 0.50

Based on 100 simulations, (0 warnings, 0 errors)
alpha = 0.05, nrow = 500

Time elapsed: 0 h 0 m 20 s
```

These results indicate that for our sample of 500 individuals nested within 25 clusters, the power of rejecting the null hypothesis of no relationship between the independent and dependent variables when the population coefficient value is 0.5, is 100%. In other words, we are almost certain to obtain a statistically significant result in this case with the specified sample sizes.

How would these power results change if our sample size was only 100 (10 individuals nested within each of 10 clusters)? The following R code will help us to answer this question.

```
x1 <- 1:10
cluster <- letters[1:10]

sim_data <- expand.grid(x1=x1, cluster=cluster)

b <- c(1, 0.5) # fixed intercept and slope
V1 <- 0.25 # random intercept variance
```

```
V2 <- matrix(c(0.5,0.3,0.3,0.1), 2) # random intercept and
slope variance-covariance matrix
s <- 1 # residual standard deviation

sim_model1 <- makeLmer(y ~ x1 + (1|cluster), fixef=b,
VarCorr=V1, sigma=s, data=sim_data)
sim_model1

powerSim(sim_model1, nsim=100)

Power for predictor 'x1', (95% confidence interval):
      100.0% (96.38, 100.0)

Test: Kenward Roger (package pbkrtest)
      Effect size for x1 is 0.50

Based on 100 simulations, (0 warnings, 0 errors)
alpha = 0.05, nrow = 100

Time elapsed: 0 h 0 m 13 s
```

Clearly, even for a sample size of only 100, the power remains at 100% for the slope fixed effect.

Rather than simulate the data for one sample size at a time, it is also possible using simr to simulate the data for multiple sample sizes, and then plot the results in a power curve. The following R code will take the simulated model described above (sim _ model1), and do just this (Figure 10.1).

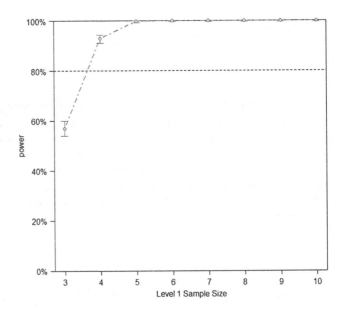

FIGURE 10.1
Power curve using simr function.

```
pc1 <- powerCurve(sim_model1)
plot(pc1, xlab="Level 1 Sample Size")
```

This curve demonstrates that, given the parameters we have specified for our model, the power for identifying the coefficient of interest as being different from 0 will exceed 0.8 when the number of individuals per cluster is 4 or more. There are a number of settings for these functions that the user can adjust in order to tailor the resulting plot to their particular needs, and we encourage the interested reader to investigate those in the software documentation. We do hope that this introduction has provided you with the tools necessary to further investigate these additional functions, however.

Summary

Our focus in Chapter 10 was on a variety of extensions to multilevel modeling. We first described models for use with unusual or difficult data situations, particularly when the normality of errors cannot be assumed, such as in the presence of outliers. We saw that there are several options available to the researcher in such instances, including estimators based on heavy-tailed distributions, as well as rank-based estimators. In practice, we might use multiple such options and compare their results in order to develop a sense as to the nature of the model parameter estimates. We then turned our attention to models designed for high-dimensional situations, in which the number of independent variables approaches (or even surpasses) the number of observations in the data set. Standard estimation algorithms will frequently yield biased parameter estimates and inaccurate standard errors in such cases. Penalized estimators such as the lasso can be used to reduce the dimensionality of the data statistically, rather than through an arbitrary selection by a researcher of predictors to retain. In addition to distributional issues, and high dimensionality, we also learned about models that are appropriate for situations in which the relationships between the independent and dependent variables are not linear in nature. There are a number of possible solutions for such a scenario, with our focus being on a spline-based solution in the form of the GAM. This modeling strategy provides the data analyst with a set of tools for selecting the optimal solution for a given dataset, and for characterizing the nonlinearity present in the data both with coefficient estimates, and graphically. We concluded the chapter with a discussion of the micro-macro modeling problem, in which level-1 variables are to serve as predictors of level-2 outcome variables, and with a review of how simulation can be used to conduct a power analysis in the multilevel modeling context. Both of these problems can be addressed using libraries in R, as we have demonstrated in this chapter.

References

Agresti, A. (2002). *Categorical Data Analysis*. Hoboken, NJ: John Wiley & Sons, Publications.

Aiken, L.S. & West, S.G. (1991). *Multiple Regression: Testing and Interpreting Interactions*. Thousand Oaks, CA: Sage.

Anscombe, F.J. (1973). Graphs in Statistical Analysis. *American Statistician, 27(1)*, 17–21.

Bickel, R. (2007). *Multilevel Analysis for Applied Research: It's Just Regression!* New York: Guilford Press.

Breslow, N. & Clayton, D.G. (1993). Approximate Inference in Generalized Linear Mixed Models. *Journal of the American Statistical Association, 88*, 9–25.

Bryk, A.S. & Raudenbush, S.W. (2002). *Hierarchical Linear Models*. Newbury Park, CA: Sage.

Bühlmann, P. & van de Geer, S. (2011). *Statistics for High-Dimensional Data: Methods, Theory and Applications*. Berlin, Germany: Springer-Verlag.

Crawley, M.J. (2013). *The R Book*. West Sussex, UK: John Wiley & Sons, Ltd.

Croon, M.A. & van Veldhoven, M.J. (2007). Predicting Group-Level Outcome Variables from Variables Measured at the Individual Level: A Latent Variable Multilevel Model. *Psychological Methods, 12(1)*, 45–57.

de Leeuw, J. & Meijer, E. (2008). *Handbook of Multilevel Analysis*. New York: Springer.

Dillane, D. (2005) Deletion Diagnostics for the Linear Mixed Model. Ph.D. Thesis, Trinity College, Dublin.

Field, A., Miles, J., & Field, Z. (2012). *Discovering Statistics Using R*. Los Angeles: Sage.

Finch, W.H. (2017). Multilevel Modeling in the Presence of Outliers: A Comparison of Robust Estimation Methods. *Psicologica, 38*, 57–92.

Fox, J. (2016). *Applied Regression Analysis and Generalized Linear Models*. Thousand Oaks, CA: Sage.

Hastie, T. & Tibshirani, R. (1990). *Generalized Additive Models*. London, UK, Chapman and Hall.

Hofmann, D.A. (2007). Issues in Multilevel Research: Theory Development, Measurement, and Analysis. In S.G. Rogelberg (Ed.). *Handbook of Research Methods in Industrial and Organizational Psychology*, pp. 247–274. Malden, MA: Blackwell.

Hogg, R.V. & Tanis, E.A. (1996). *Probability and Statistical Inference*. New York: Prentice Hall.

Hox, J. (2002). *Multilevel Analysis: Techniques and Applications*. Mahwah, NJ: Erlbaum.

Iversen, G. (1991). *Contextual Analysis*. Newbury Park, CA: Sage.

Jaekel, L.A. (1972). Estimating Regression Coefficients by Minimizing the Dispersion of Residuals. *Annals of Mathematical Statistics, 43*, 1449–1458.

Kloke, J. & McKean, J.W. (2013). Small Sample Properties of JR Estimators. Paper presented at the annual meeting of the American Statistical Association, Montreal, QC, August.

Kloke, J.D., McKean, J.W., & Rashid, M. (2009). Rank-Based Estimation and Associated Inferences for Linear Models with Cluster Correlated Errors. *Journal of the American Statistical Association, 104*, 384–390.

Kreft, I.G.G. & de Leeuw, J. (1998). *Introducing Multilevel Modeling.* Thousand Oaks, CA: Sage.

Kreft, I.G.G., de Leeuw, J., & Aiken, L. (1995). The Effect of Different Forms of Centering in Hierarchical Linear Models. *Multivariate Behavioral Research, 30,* 1–22.

Kruschke, J.K. (2011). *Doing Bayesian Data Analysis.* Amsterdam, Netherlands: Elsevier.

Lange, K.L., Little, R.J.A., & Taylor, J.M.G. (1989). Robust Statistical Modeling Using the *t* Distribution. *Journal of the American Statistical Association, 84,* 881–896.

Liu, Q. & Pierce, D.A. A Note on Gauss-Hermite Quadrature (1994). *Biometrika, 81,* 624–629.

Lynch, S.M. (2010). *Introduction to Applied Bayesian Statistics and Estimation for Social Scientists.* New York: Springer.

Pinheiro, J., Liu, C., & Wu, Y.N. (2001). Efficient Algorithms for Robust Estimation in Linear Mixed-Effects Models Using the Multivariate t Distribution. *Journal of Computational and Graphical Statistics, 10,* 249–276.

Potthoff, R.F. & Roy, S.N. A Generalized Multivariate Analysis of Variance Model Useful Especially for Growth Curve Problems. *Biometrika, 51(3–4),* 313–326.

R Development Core Team (2012). *R: A Language and Environment for Statistical Computing.* Vienna, Austria: R Foundation for Statistical Computing.

Rogers, W.H. & Tukey, J.W. (1972). Understanding Some Long-Tailed Symmetrical Distributions. *Statistica Neerlandica, 26(3),* 211–226.

Sarkar, D. (2008). *Lattice: Multivariate Data Visualization with R.* New York: Springer.

Schall, R. (1991). Estimation in Generalized Linear Models with Random Effects. *Biometrika, 78,* 719–727.

Schelldorfer, J., Bühlmann, P., & van de Geer, S. (2011). Estimation for High-Dimensional Linear Mixed-Effects Models Using l1-Penalization. *Scandinavian Journal of Statistics, 38,* 197–214.

Snijders, T. & Bosker, R. (1999). *Multilevel Analysis: An Introduction to Basic and Advanced Multilevel Modeling,* 1st edition. Thousand Oaks, CA: Sage.

Snijders, T. & Bosker, R. (2012). *Multilevel Analysis: An Introduction to Basic and Advanced Multilevel Modeling,* 2nd edition. Thousand Oaks, CA: Sage.

Song, P.X.-K., Zhang, P., & Qu, A. (2007). Maximum Likelihood Inference in Robust Linear Mixed-Effects Models Using Multivariate *t* Distributions. *Statistica Sinica, 17,* 929–943.

Staudenmayer, J., Lake, E.E., & Wand, M.P. (2009). Robustness for General Design Mixed Models Using the t-Distribution. *Statistical Modeling, 9,* 235–255.

Tibshirani, R. (1996). Regression Shrinkage and Selection via the Lasso. *Journal of the Royal Statistical Society, Series B, 58,* 267–288.

Tong, X. & Zhang, Z. (2012). Diagnostics of Robust Growth Curve Modeling Using Student's *t* Distribution. *Multivariate Behavioral Research, 47(4),* 493–518.

Tu, Y.-K., Gunnell, D., & Gilthorpe, M.S. (2008). Simpson's Paradox, Lord's Paradox, and Suppression Effects are the Same Phenomenon – The Reversal Paradox. *Emerging Themes in Epidemiology, 5(2),* 1–9.

Tukey, J.W. (1949). Comparing Individual Means in the Analysis of Variance. *Biometrics, 5(2),* 99–114.

Wang, Y. (1998). Mixed Effects Smoothing Spline Analysis of Variance. *Journal of the Royal Statistical Society B, 60(1),* 159–174.

Wang, J. & Genton, M.G. (2006). The Multivariate Skew-Slash Distribution. *Journal of Statistical Planning and Inference, 136,* 209–220.

Welsh, A.H. & Richardson, A.M. (1997). Approaches to the Robust Estimation of Mixed Models. In G. Maddala and C.R. Rao (Eds.). *Handbook of Statistics, vol. 15,* pp. 343–384. Amsterdam: Elsevier Science B.V.

Wilcoxon, F. (1945). Individual Comparisons by Ranking Methods. *Biometrics Bulletin, 1(6),* 80–83.

Wolfinger, R. & O'Connell, M. (1993). Generalized Linear Mixed Models: A Pseudo-Likelihood Approach. *Journal of Statistical Computation and Simulation, 48,* 233–243.

Wood, S.N. (2006). *Generalized Additive Models: An Introduction with R.* New York: Chapman and Hall/CRC.

Wooldridge, J. (2004). *Fixed Effects and Related Estimators for Correlated Random Coefficient and Treatment Effect Panel Data Models.* East Lansing: Department of Economics, Michigan State University.

Yuan, K.-H. & Bentler, P.M. (1998). Structural Equation Modeling with Robust Covariances. *Sociological Methodology, 28,* 363–396.

Yuan, K.-H., Bentler, P.M., & Chan, W. (2004). Structural Equation Modeling with Heavy Tailed Distributions. *Psychometrika, 69(3),* 421–436.

Zhao, P. & Yu, B. (2006). On Model Selection Consistency of Lasso. *Journal of Machine Learning Research, 7,* 2541–2563.

Index

Printed in the United States
by Baker & Taylor Publisher Services